PRAISE FOR *THE FABRIC OF CIVILIZATION*

"*The Fabric of Civilization: How Textiles Made the World* is one of the most compelling books I've ever read on the history of textiles—witty, well-researched, and full of fascinating things I'd never known despite being quite well-read on textile technology. (Did you know that letters of credit from clothiers helped build the early European banking system? Or that aniline dyes were one of the earliest applications of organic chemistry?) If you thought Elizabeth Wayland Barber's *Women's Work: The First 20,000 Years* was interesting, you'll swoon over this book. I *loved* it."

—Tien Chiu, weaver, teacher, and author of *Master Your Craft*

"Postrel's brilliant, learned, addictive book tells a story of human ingenuity. . . . Her deep story is of the liberty that permitted progress. Presently the descendants of slaves and serfs and textile workers got closets full of beauty, and fabric for the cold, a Great Enrichment since 1800 of three thousand percent."

—Deirdre Nansen McCloskey, author of the Bourgeois Era trilogy

"Cleanly written and completely accessible, this book opens up an entirely new world of textiles, explaining the most ancient archeological fabrics and the latest polymer blends that cool the body—not warm it as textiles have done for thousands of years—with equal verve."

—Valerie Hansen, author of *The Year 1000*

"*The Fabric of Civilization* is more engaging and informative than any textile science or textile history course that I studied in college. Postrel has distilled thousands of years of the making and manipulation of string (thread) into a comprehensible read whether or not you have knowledge of its invention or the current bio-engineering research. She has woven in personal experiences; interviewed historians, computer programmers, scientists, textile archeologists, and contemporary and indigenous artisans; and gone behind the scenes in textile laboratories and manufacturing plants. After reading this book, I felt like I had wandered the world for centuries using fabric for my currency and wealth. I had won political battles in its name. I had been sickened as well as healed by it. And now I am an inventor synthesizing environmentally healthy fabric by means of chemicals. My appreciation and knowledge of how textiles make the world have been greatly enhanced."

—Marilyn Murphy, cofounder ClothRoads and former president of Interweave

"Fiber artists of any stripe, prepare for a juicy read through a new book on textile history! Virginia Postrel's new book will turn your old ideas about fabric and civilization down unexpected pathways. *The Fabric of Civilization* is divided into chapters loosely related to individual textile processes, such as spinning, weaving, and dyeing, and the author points out that textiles are the basis of our number systems, our banking, our commerce, and our science. This book is full of stories of individuals who innovated entire systems because of their sensitivity to textile processes. The author actually learned to spin and weave while writing this book, and her explanations and diagrams are spot on. Profusely illustrated with photos and diagrams, and contains a comprehensive bibliography. Highly recommended."

—Alice Schlein, weaver and author of *Network Drafting* and *The Woven Pixel*

"*The Fabric of Civilization: How Textiles Made the World* chronicles the laborious and cumulative innovations that allow cloth to play an essential role in our comfort, cultural identity, and our dependence on programmable functions. At times, the fabric of society appears threadbare but, based on the global nature of textiles, it is comforting to know that all cultures have a shared experience. Postrel reminds us that we are all woven together with colorful threads."

—Susie Taylor, artist

THE
FABRIC OF
CIVILIZATION

THE
FABRIC OF
CIVILIZATION

HOW TEXTILES MADE
THE WORLD

VIRGINIA POSTREL

BASIC BOOKS

New York

Basic Books
Hachette Book Group
1290 Avenue of the Americas, New York, NY 10104
www.basicbooks.com

Printed in the United States of America

First Edition: December 2020

Published by Basic Books, an imprint of Perseus Books, LLC, a subsidiary of Hachette Book Group, Inc. The Basic Books name and logo is a trademark of the Hachette Book Group.

The Hachette Speakers Bureau provides a wide range of authors for speaking events. To find out more, go to www.hachettespeakersbureau.com or call (866) 376-6591.

The publisher is not responsible for websites (or their content) that are not owned by the publisher.

Print book interior design by Amy Quinn

Library of Congress Cataloging-in-Publication Data
Names: Postrel, Virginia I., 1960– author.
Title: The fabric of civilization : how textiles made the world / Virginia Postrel.
Description: First edition. | New York : Basic Books, 2020. | Includes bibliographical references and index.
Identifiers: LCCN 2020014544 | ISBN 9781541617605 (hardcover) | ISBN 9781541617612 (ebook)
Subjects: LCSH: Textile industry—History. | Textile fabrics—History.
Classification: LCC HD9850.5 .P67 2020 | DDC 338.4/7677009—dc23
LC record available at https://lccn.loc.gov/2020014544ISBNs: 978-1-5416-1760-5 (hardcover), 978-1-5416-1761-2 (ebook)

LSC-C

Printing 5, 2021

To my parents,
Sam and Sue Inman,
and
to Steven

CONTENTS

THE
FABRIC OF
CIVILIZATION

THE FABRIC OF CIVILIZATION

The most profound technologies are those that disappear.
They weave themselves into the fabric of everyday
life until they are indistinguishable from it.

—Mark Weiser, "The Computer for the 21st
Century," *Scientific American*, September 1991

IN 1900, A BRITISH ARCHAEOLOGIST made one of the greatest finds of all time. Arthur Evans, later knighted for his discoveries, unearthed the palace complex at Knossos on Crete. With its intricate architecture and gloriously painted frescoes, the site bore witness to a sophisticated Bronze Age civilization, more ancient than anything found on the Greek mainland. A scientist with a classical education and a poetic streak, Evans named the vanished inhabitants Minoans. In Greek legend, Minos, the first king of Crete, demanded that every nine years the Athenians send seven boys and seven girls to be sacrificed to the Minotaur.

"Here," Evans wrote in a newspaper article, "Daedalus constructed the Labyrinth, the den of the Minotaur, and fashioned the wings—perhaps the sails—with which he and Icarus took flight over the Aegean." At Knossos, too, the Athenian hero Theseus had unwound a ball of thread as he journeyed through the Labyrinth, killed the ferocious man-bull, and followed the thread back to freedom.

Like Troy before it, the city of legends turned out to be real. Excavations revealed a literate and well-organized civilization as old as those of Babylon and Egypt. The find also presented a linguistic mystery. Along with art, pottery, and ritual objects, Evans uncovered thousands of clay tablets inscribed with characters he'd seen in the artifacts that had drawn him to Crete in the first place. He identified two distinct scripts, along with hieroglyphs representing objects such as a bull's head, a spouted vase, and what Evans took to be a palace or tower: a rectangle bisected on the diagonal, with four spikes on top. But he couldn't read the tablets.

Although he worked on the problem for decades, Evans never managed to decipher them. Not until 1952, eleven years after his death, was one script finally identified as an early form of Greek. Much of the other script remains unreadable. But we do know that Evans got his "tower" upside down and completely mistook its meaning. The hieroglyph depicted not a crenellated battlement but a fringed piece of fabric or perhaps a warp-weighted loom. It meant not *palace*, but *textile*.

The Minoan culture that inspired stories of lifesaving thread kept meticulous accounts of large-scale wool and flax production. Textile records make up more than half of the tablets ultimately uncovered at Knossos. They track "textile crops, the birth of lambs, targets for wool yields per animal, collectors' work, the assignment of wool to workers, the receipt of finished fabrics, distribution of cloth or clothing to dependent personnel, and the storage of cloth in the palatial magazines," writes a historian. In a single season, palace workshops processed fleece from seventy to eighty thousand sheep, spinning and weaving an astonishing sixty tons of wool.

Evans had missed the source of the city's wealth and the primary activity of its residents. Knossos was a textile superpower. Like many people before and since, the pioneering archaeologist had overlooked the central role of textiles in the history of technology, commerce, and civilization itself.[1]

We hairless apes coevolved with our cloth. From the moment we're wrapped in a blanket at birth, we are surrounded by textiles. They cover our bodies, bedeck our beds, and carpet our floors. Textiles give us seat belts and sofa cushions, tents and bath towels, medical masks and duct tape. They are everywhere.

But, to reverse Arthur C. Clarke's famous adage about magic, any sufficiently familiar technology is indistinguishable from nature.[2] It seems intuitive, obvious—so woven into the fabric of our lives that we take it for granted. We no more imagine a world without cloth than one without sunlight or rain.

We drag out heirloom metaphors—"on tenterhooks," "towheaded," "frazzled"—with no idea that we're talking about fabric and fibers. We repeat threadbare clichés: "whole cloth," "hanging by a thread," "dyed in the wool." We catch airline shuttles, weave through traffic, follow comment threads. We speak of life spans and spinoffs and never wonder why drawing out fibers and twirling them into thread looms so large in our language. Surrounded by textiles, we're largely oblivious to their existence and to the knowledge and efforts embodied in every scrap of fabric.

Yet the story of textiles is the story of human ingenuity.

Agriculture developed in pursuit of fiber as well as food. Labor-saving machines, including those of the Industrial Revolution, came out of the need for thread. The origins of chemistry lie in the coloring and finishing of cloth; the beginning of binary code—and aspects of mathematics itself—in weaving. As much as for spices or gold, the quest for fabrics and dyestuffs drew merchants to cross continents and sailors to explore strange seas.

From the most ancient times to the present, the textile trade has fostered long-distance exchange. The Minoans exported woolen cloth, some of it dyed in precious purple, as far away as Egypt. The ancient Romans wore Chinese silk, worth its weight in gold. The textile business funded the Italian Renaissance and the Mughal Empire; it left us Michelangelo's *David* and the Taj Mahal. It spread the alphabet and double-entry bookkeeping, gave rise to financial institutions, and nurtured the slave trade.

In ways both subtle and obvious, beautiful and terrible, textiles made our world.

The global story of textiles illuminates the nature of civilization itself. I use this term not to imply moral superiority or the end state of an inevitable progression but in the more neutral sense suggested by this definition: "the accumulation of knowledge, skills, tools, arts, literatures, laws, religions and philosophies which stands between man and external nature, and which serves as a bulwark again the hostility of forces that would otherwise destroy him."[3] This description captures two critical dimensions that together distinguish civilization from related concepts, such as culture.

First, civilization is *cumulative*. It exists in time, with today's version built on previous ones. A civilization ceases to exist when that continuity is broken. Minoan civilization disappeared. Conversely, a civilization may evolve over a long stretch of time while the cultures that make it up pass away or change irrevocably. The western Europe of 1980 was radically different in its social mores, religious practices, material culture, political organization, technological resources, and scientific understanding from the Christendom of 1480, yet we recognize both as Western civilization.

The story of textiles demonstrates this cumulative quality. It lets us trace the progress and interactions of practical techniques and scientific theory: the cultivation of plants and breeding of animals, the spread of mechanical innovations and measurement standards, the recording and replication of patterns, the manipulation of chemicals. We can watch knowledge spread from one place to another, sometimes in written form but more often through human contact or the exchange of goods, and see civilizations become intertwined.

Second, civilization is a *survival technology*. It comprises the many artifacts—designed or evolved, tangible or intangible—that stand between vulnerable human beings and natural threats, and that invest the world with meaning. Providing protection and adornment, textiles are themselves among such artifacts. So, too, are the innovations they inspire, from better seeds to weaving patterns to new ways of recording information.

Along with the perils and discomforts of indifferent nature, civilization protects us from the dangers posed by other humans. Ideally, it allows us to live in harmony. Eighteenth-century thinkers used the term to refer to the intellectual and artistic refinement, sociability, and peaceful interactions of the commercial city.[4] But rare is the civilization that exists without organized violence. At best, civilization encourages cooperation, curbing humanity's violent urges; at worst, it unleashes them to conquer, pillage, and enslave. The history of textiles reveals both aspects.

It also reminds us that technology means much more than electronics or machines. The ancient Greeks worshiped Athena as the goddess of *techne*: craft and productive knowledge, the artifice of civilization. She was the giver and protector of olive trees, of ships, and of weaving. The Greeks used the same word for two of their most important technologies, calling both the loom and the ship's mast *histós*. From the same root, they dubbed sails *histía*, literally the product of the loom.[5]

To weave is to devise, to invent—to contrive function and beauty from the simplest of elements. In *The Odyssey*, when Athena and Odysseus scheme, they "weave a plan." *Fabric* and *fabricate* share a common Latin root, *fabrica*: "something skillfully produced." *Text* and *textile* are similarly related, from the verb *texere*, "to weave," which in turn derives—as does *techne*—from the Indo-European word *teks*, meaning "to weave." *Order* comes from the Latin word for setting warp threads, *ordior*, as does the French word for computer, *ordinateur*. The French word *métier*, meaning a trade or craft, is also the word for loom.

Such associations aren't uniquely European. In the K'iche' Maya language, the terms for weaving designs and writing hieroglyphics both use the root *-tz'iba-*. The Sanskrit word *sutra*, which now refers to a literary aphorism or religious scripture, originally denoted string or thread; the word *tantra*, which refers to a Hindu or Buddhist religious text, is from the Sanskrit *tantrum*, meaning "warp" or "loom." The Chinese word *zuzhi*, meaning "organization" or "arrange," is also the word for weave, while *chengji*, meaning "achievement" or "result," originally meant twisting fibers together.[6]

Cloth making is a creative act, analogous to other creative acts. It is a sign of mastery and refinement. "Can we expect, that a government will be well modelled by a people, who know not how to make a spinning-wheel, or to employ a loom to advantage?" wrote philosopher David Hume in 1742.[7] The knowledge is all but universal. Rare is the people that does not spin or weave, and rare, too, the society that does not engage in textile-related trade.

The story of textiles is a story of famous scientists and forgotten peasants, incremental improvements and sudden leaps, repeated inventions and once-ever discoveries. It is a story driven by curiosity, practicality, generosity, and greed. It is a story of art and science, women and men, serendipity and planning, peaceful trade and savage wars. It is, in short, the story of humanity itself—a global story, set in every time and place.

Like the carefully planned *strip cloths* of West Africa, *The Fabric of Civilization* is a whole made up of distinct pieces, each interwoven, with its own *warp* and *weft*.[8] Each chapter's warp represents a stage in the textile journey. We begin with production—fiber, thread, cloth, dye—and then move, like cloth itself, to merchants and consumers. Finally, we return to a new take on fiber, meeting the innovators who revolutionized textiles in the twentieth century as well as some today who hope to use cloth to change the world. Within each chapter, events take place in roughly chronological order. Think of the warp as the chapter's *what*.

The weft constitutes the *why*—some significant influence of textile materials, makers, or markets on the character and progress of civilization. We explore the artifice behind "natural" fibers and discover why spinning machines set off an economic revolution. We examine the deep relationship between cloth and mathematics and what dye tells us about chemical knowledge. We look at the essential role of "social technologies" in enabling trade, the many ways in which the desire for textiles disrupts the world, and the reasons textile research appeals even to pure scientists. The weft supplies the broader context for the chapter's history.

Each chapter can be read separately, just as a single strip of *kente cloth* can form a stole. But the whole reveals the greater pattern. From prehistory to the near future, this is the story of the human beings who wove, and are still weaving, civilization's tale.

Chapter One

FIBER

The Lord is my shepherd. I shall not want.

—Psalm 23

IN THESE DAYS OF SPANDEX blends and performance microfibers, Levi's still sells some old-fashioned 100 percent cotton jeans. Look closely and you can see the structure. Each thread is fine, long, and even, extending the full length or breadth of the garment. The vertical threads are blue with a white core, while the horizontal ones, revealed in the artfully placed rips, are white all the way through. In worn areas and on the inside, you can see the diagonal pattern of the twill weave that gives denim its durability and natural stretch.

We call cotton a "natural fiber," a value-laden contrast to synthetics like polyester and nylon. But it is nothing of the sort. Thread, dye, cloth, even the plants and animals that supply the raw material, are all the products of millennia of refinements and innovations, large and small. Human action, not nature alone, made cotton what it is today.

Cotton, wool, linen, silk, and their less prominent kin may have biological origins, but these so-called natural fibers are the products of artifice so ancient and familiar that we forget it's there. The journey to finished cloth begins with plants and animals bred by trial and error to produce unnaturally abundant fiber suitable for making thread. These genetically modified organisms are technological achievements every bit as ingenious as the machines we honor as the Industrial Revolution. And they, too, have far-reaching consequences for economics, politics, and culture.

⚏⚏⚏⚏

What we usually call the Stone Age could just as easily be called the String Age. The two prehistoric technologies were literally intertwined. Early humans used string to attach stone blades to handles, creating axes and spears.

The blades survived the millennia, waiting to be excavated by archaeologists. The cords rotted away, their vestiges invisible to the naked eye. Scholars named prehistoric ages after the layers of increasingly sophisticated stone tools they found: Paleolithic, Mesolithic, Neolithic. *Lithic* means "of or pertaining to stone." Nobody thought about the missing threads. But we get a false picture of prehistoric life and of the earliest products of human ingenuity when we imagine only the hard tools that easily endure the passage of time. Today's researchers can detect the traces of softer stuff.

Bruce Hardy, a paleoanthropologist at Kenyon College in Ohio, specializes in what is known as residue analysis—looking at the microscopic fragments left behind when the earliest stone tools cut through other materials. To build a library of comparison samples, he uses replicas to chop up plants and animals that early people might have used, then examines the tools under a microscope. By learning their microscopic characteristics, he can identify tuber cells and mushroom spores, fish scales and feather fragments. And he can spot fibers.

In 2018, he was working in the Paris lab of Marie-Hélène Moncel, examining tools she'd excavated from a site in southeastern France called Abri du Maras. There, about forty to fifty thousand years ago, Neanderthal people lived under the protection of an overhanging rock shelter. Three meters below today's surface, they left a layer containing ashes, bones, and stone tools. Hardy had previously found individual twisted plant fibers on some of their tools—tantalizing evidence suggesting that they might have made string. But a single fiber is not a cord.

This time, Hardy spotted a pimple-sized bit of cream on a two-inch stone tool. Easily overlooked on the sand-colored surface of the flint, to his practiced eye it could have been a neon sign blinking THIS IS IT! "As soon as I saw it, I knew there was something else going on there," he says. "I was thinking, 'Wow, this is it. I think we got it now.'" Caught in the stone was a bunch of twisted fibers.

As Hardy and his colleagues examined the find with increasingly sensitive microscopes, it got even more exciting. Three distinct bundles of fibers, each twisted in the same direction, had been twisted together in the opposite direction to form a three-ply cord. Using fibers from the inner bark of conifer trees, Neanderthal people had made string.

Like the steam engine or the semiconductor, string is a general-purpose technology with countless applications. With it, early humans could create fishing lines and nets, make bows for hunting or starting fires, set traps for small game, wrap and carry bundles, hang food to dry, strap babies to their chests, fashion belts and necklaces, and sew together hides. String expanded the capabilities of human hands and built the capacity of human minds.

"As the structure becomes more complex (multiple cords twisted to form rope, ropes interlaced to form knots)," Hardy and his coauthors write, "it demonstrates an 'infinite use of finite means' and requires a cognitive complexity similar to that required by human language." Whether used to fashion snares or to tie bundles, string made catching, carrying, and storing provisions easier. It gave early hunter-gatherers more flexibility and control over their environment. Its invention was a fundamental step toward civilization.

"So powerful, in fact, is simple string in taming the world to human will and ingenuity that I suspect it to be the unseen weapon that allowed the human race to conquer the earth," writes textile historian Elizabeth Wayland Barber.[1] Our distant forebears may have been primitive, but they were also clever and inventive. They left behind striking artworks and world-altering technologies: cave paintings, small sculptures, bone flutes, beads, bone needles, and compound tools, including detachable spearheads and harpoon points. Although string survived the millennia only in trace amounts, it was part of the same creative profusion.

The earliest sources were *bast fibers*, which grow just inside the bark of trees and the outer stem of such plants as flax, hemp, ramie, nettle, and jute. Tree fibers tend to be coarser and take more effort to extract. Plus, notes Hardy, "it takes a lot less time for flax to grow than it does a tree."

Discovering how to harvest fiber from wild flax thus represented a significant advance. It's easy to imagine how it might have happened. When stems fell on the ground, the outer layers rotted away in the dew or rain, exposing the long, stringy strands within. Early humans could strip away the fibers and twist them into string, rolling the flax between their fingers or against their thighs.

Whether from slow-growing trees or fast-growing plants, bast fibers alone didn't make string abundant. When the only way to create cord is to roll bast fibers on your thigh, making enough to create a looped bag can take the equivalent of two modern workweeks, between 60 and 80 hours, based on traditional practices in Papua New Guinea. Looping the bag into shape can require another 100 to 160 hours—a month's labor.[2]

𝕫𝕫𝕫𝕫

String may be a powerful technology, but it is not cloth. To produce enough thread to make fabric, you need a larger, more predictable supply of raw material. You need fields of flax, flocks of sheep, and the time to transform disordered masses of fiber into many yards of thread. You need agriculture—a technological leap that quickly expanded from food to fibers.

It's called the Neolithic Revolution. Roughly twelve thousand years ago, humans began to establish permanent settlements and to cultivate plants and domesticate animals. Though they continued to hunt and forage, these people no longer subsisted solely on what they found in their environment. By understanding and controlling reproduction, they began to alter plants and animals to suit their own purposes. Along with new sources of food, they invented "natural" fibers.

Eleven thousand years ago, somewhere in southwest Asia, sheep joined dogs as the earliest domesticated animals. These Neolithic sheep weren't the woolly white creatures of Nativity scenes, mattress ads, or Australian pastures. Their coats were brown, with coarse hair that molted each spring,

A primitive Soay sheep, the closest living relative of sheep before human breeding. Note the molting fleece. Compare to the modern Merino sheep. (iStockphoto)

falling out in clumps rather than growing continuously. The early herders slaughtered most males and many females while the animals were still young, using them for meat. They allowed only those with the most desirable qualities to mature and breed. Over time—a very, very long time—human choices changed the nature of sheep. The animals got shorter, their horns shrank, their coats grew increasingly woolly, and, although ancient shepherds plucked rather than sheared their animals, domesticated sheep eventually stopped molting.

After about two thousand generations—more than five thousand years, or halfway to the present day—selective breeding had transformed sheep into the wool-producing creatures depicted in Mesopotamian and Egyptian art. They had thick fleeces in a range of colors, including white, and thicker bones to support their heavier coats. Over time, the fibers of their fleece became finer and more uniform. Bone excavations show that the mix of flocks also changed. In earlier sites archaeologists find almost exclusively bones from lambs slaughtered for meat, whereas in later ones many bones also come from sheep that had survived to adulthood, including (likely castrated) males. Ancient people had begun producing wool.[3]

Something similar happened with the grassy wildflower known as flax. In the wild, flax seedpods burst open when ripe and drop their tiny seeds on the ground, where they're nearly impossible to collect. Early farmers collected the pods from the rare plants on which they remained closed. Like blue eyes, these intact capsules express a recessive genetic trait, so seeds from them produce offspring whose seedpods also stay closed. Most of the harvested seeds were either eaten or pressed for oil, but cultivators held back the largest to plant the next season. Over time, domesticated flax seeds grew bigger than their wild kin, providing more of the oil and nutrients humans valued.[4]

Agricultural pioneers then created a second kind of domesticated flax. They preserved seeds from taller plants with fewer branches and pods. In these, the plants' energy went into their stems, yielding more fiber. Fields of this flax could supply enough material to make linen cloth.[5]

But merely growing flax plants doesn't produce thread suitable for weaving. First, the fiber has to be harvested and processed—an elaborate business, even today. The first step is to pull up each stalk by the roots, preserving the full length of fiber. You then allow the harvested stalks to dry. Next comes a smelly process called *retting*, where the stems are kept soaked in water so that bacteria break down the sticky pectin that holds the useful fibers to the inner stem. Unless the water is a free-flowing stream, retting stinks to high heaven. The similarity between *ret* and *rot* isn't a coincidence.

A woman dreams of magical relief from the arduous labor of processing flax in this anonymous Dutch print from around 1673. (Rijksmuseum)

Judging the right time to take the stalks out of the water is tricky. Too soon and the fibers will be hard to remove, not soon enough and they'll break into tiny pieces. Once out of the water, the stalks have to dry thoroughly before being beaten and scraped to separate the fibers from the straw, a step called *scutching*. Finally comes *hackling*, in which the fibers are run through combs to separate the long fibers from the short, fluffy tow. Only then is the flax ready to spin into thread.

Given all this effort, early humans clearly put a high value on linen. We don't know precisely when people started cultivating flax to produce cloth rather than oil, but we do know that it must have happened in the earliest days of agriculture. In 1983, archaeologists working in the Nahal Hemar Cave, near the Dead Sea in Israel's Judaean Desert, found scraps of linen yarn and fabric, including the remains of what appears to be some type of headgear. Radiocarbon-dated to nearly nine thousand years ago, these textiles predate pottery and may even predate looms. Rather than woven, the cloth was made with twining, knotting, and looping techniques akin to those used in basketry, macramé, or crochet.

The cave's textiles weren't rudimentary experiments but the work of skilled artisans who clearly knew what they were doing. The remains reveal techniques that require time to perfect. An archaeologist analyzing them cites "the fine workmanship, the degree of regularity and delicacy, the sophisticated details and the keen sense for decoration exhibited. Finishing touches included sewing with buttonhole and 'stroke' stitches," evenly spaced, parallel embroidery stitches of the same length. The thread is strong and smoothly spun, not what you would get by stripping fiber from random stalks on the ground and twisting it together with your fingers. In some cases two spun strands have been plied together for strength.[6]

Nine thousand years ago, in other words, Neolithic farmers had already figured out not only how to breed and grow flax for fiber but also how to process and spin it into high-quality thread and how to turn that thread into decoratively stitched cloth. Textiles date to the earliest days of permanent settlements and agriculture.

Transforming sheep and flax into reliable sources of raw material for thread took careful observation, ingenuity, and patience. But that was nothing compared to the imagination—and genetic good fortune—required to turn cotton into the world's dominant, and most historically consequential, "natural" fiber.

◰◰◰◰

Suspended from the branches a foot or so about my head are what look like cocoons, with shadowy cores visible through wispy fibers. One dangles like a fluffy white spider from a three-inch strand. When I pluck it, the thread is soft and slightly twisted, not at all like sticky cocoon silk. The dark core is a hard seed. This is cotton, *Gossypium hirsutum*, from the Yucatan Peninsula, the wild version of today's dominant commercial species. Looking at the little thread, stretched and twisted by nature, I can see how early humans got the idea that these filaments might prove useful.

"It's forms like this that at least four separate times, by four separate cultures—in each case extending back five thousand years or more—first attracted the attention of aboriginal domesticators," says evolutionary biologist Jonathan Wendel. "They brought it into domestication slowly but surely, using it for seed oil, to feed their domesticated animals, or because they were making lamp wicks, pillow stuffing, wound dressing—this incredible diversity of uses."

We're in the greenhouse atop a building at Iowa State University, the unlikely Corn Belt home of one of the world's leading experts on cotton genetics—and one of the most dedicated collectors and cultivators of rare specimens. The greenhouse shelters hundreds of cotton plants, representing about twenty different species from all over the world, along with samples of *Gossypium*'s closest relatives: Hawaii's *Kokia* and Madagascar's *Gossypioides*. Cotton gets around. "All these plants have stories," says Wendel, a lean marathon runner who exudes a contagious enthusiasm for the weird natural history of cotton.

Most of the world's fifty or so wild cotton species are completely useless for making thread. Their seeds carry no more fuzz than a peach. But roughly a million years ago, the seeds of one African *Gossypium* species began to sprout longer bits of fluff, each fiber a single, twisted cell. "This happened only once, in this African group," Wendel says.

In his office, he hands me a plastic bag of tiny wild *Gossypium herbaceum* bolls, from the closest surviving descendant of the African species from which all cotton fiber comes. They're mostly seeds, with just enough fluff to hang together. "Long before there were humans, nature gave us that," he says. Scientists aren't sure why the fiber evolved. It doesn't serve to attract birds, which only rarely disperse cotton seeds anyway. Perhaps it helps the seeds germinate, by attracting microbes that break down the tough seed coat when enough water is present. We just don't know. Whatever the reason, a distinctive fiber-producing cotton genome survived. Scientists call it the A genome.

The fiber-producing mutation was the first lucky break for future denim wearers. Not long afterward, something even more remarkable happened. An African cotton seed somehow crossed the ocean to Mexico. It took root and crossbred with a local species that had evolved its own distinctive genome, called D. Like the rest of the world's cotton species, the D cotton produced no fiber, but the new hybrid did. In fact, it had the genetic potential to develop varieties with even more fiber than its African parent. That's because instead of getting the usual single copy of each parent's chromosomes, it got both copies, giving it twenty-six pairs of chromosomes to work with rather than thirteen. (This phenomenon, called *polyploidy* as opposed to the normal *diploidy*, is common in plants.) Geneticists refer to the New World hybrid as AD.

Like the original African mutation, the transoceanic AD hybrid occurred only once. When Wendel started working on cotton in the 1980s, there were two competing theories of how the A and D genomes got together. The first was that the hybrid emerged at least sixty-five million years ago, when South America and Africa were still part of a single landmass, before a shift in the earth's tectonic plates led the two continents to drift apart. "At the other end of the spectrum," he recalls, "were Kon-Tikiists," who argued that humans must have brought seeds with them on boats, so "polyploid cottons were maybe five thousand or ten thousand years old." (*Kon-Tiki* was a balsa raft used by Thor Heyerdahl to sail from Peru to French Polynesia in 1947, testing the hypothesis that ancient people could have made long-distance sea voyages.)

Both were wrong. Geneticists can now estimate the age of a species by sequencing its DNA to see how much the component base pairs differ from those of a related species. Mutations take place at a reasonably predictable rate that can be calibrated with fossil evidence to indicate when the two species diverged from a common ancestor. The mutation rate varies—in the plant world, trees change more slowly than annuals, for instance—and not every species has a fossil record, so estimates aren't precise. But they get you in the ballpark. "You can be off by a multiple of two or three or four," says Wendel, "but not by ten or a hundred or a thousand."

In the case of the mysterious cotton hybrid, that's good enough. The parent A and D genomes and the AD hybrid are way too similar for the merger to date all the way back to when dinosaurs walked the earth—A and D themselves only diverged between five and ten million years ago—and they're much too different for the hybrid to be the product of human transport. "There's not a snowball's chance in hell it was *Kon-Tiki*," says Wendel. "Polyploid cotton certainly formed before humans were walking on the planet."

We have no idea how the cotton seed managed to cross the ocean, or even whether it went west across the Atlantic or east across the Pacific. Maybe it floated on a piece of pumice or got caught up in a hurricane. Whatever it was, something extremely unlikely happened. "It's the evolutionary significance of really rare events," says Wendel.

In this case, the significance was not only evolutionary but commercial and cultural. Once people arrived, the extra genetic complement gave cultivators in the Americas many more possibilities to work with. As a result, says Wendel, "human selection was able to create longer, stronger, finer cotton than was possible in the A-genome domesticates in the Old World." With its AD genome, New World cotton, the ancestor of the species that fed the Industrial Revolution and gave us blue jeans, owes its existence to an amazing stroke of luck.

In its natural state, however, even the most fruitful cotton plant would be a sorry source for making thread, much less cloth. On both sides of the Atlantic, wild cotton is a scarce and scraggly shrub. Its small bolls are mostly seeds whose coats are so tough they rarely germinate. Long before anyone conceived the term *genetically modified organism*, human action turned this

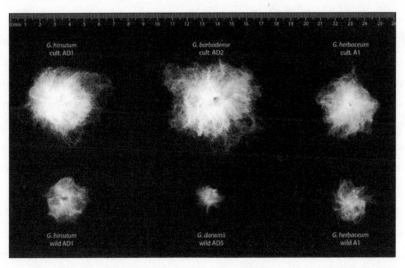

Domesticated cotton has longer, whiter, more plentiful fibers than wild cotton. (Jonathan Wendel)

unpromising plant into what Wendel calls a "fruit machine." People created the fiber-laden bolls we now know as cotton.

In southern Africa and the Indus Valley, on the Yucatan Peninsula and the coast of Peru, farmers saved the seeds from the plants with longer, more abundant fibers for future crops. They learned to nick the tough seed coats to encourage sprouting and to look for seeds that weren't quite as hard. They favored white bolls over nature's palette of browns. They rewarded plants that ripened quickly and at about the same time. From this biological manipulation came four domesticated cotton species: two in the Old World, *Gossypium arboreum* and *Gossypium herbaceum*, and two in the New, *Gossypium hirsutum* and *Gossypium barbadense*.

"The four cotton species," write Wendel and his coauthors in an overview of cotton domestication, "were transformed from undisciplined perennial shrubs and small trees with small impermeable seeds sparsely covered by coarse, poorly differentiated seed hairs, to short, compact, annualized plants with copious amounts of long, white lint borne on large seeds that germinate readily."[7]

So far so good. But for thousands of years, many of today's most important cotton-growing regions wouldn't support any of the four domesticated species. You couldn't raise cotton in the Mississippi delta, on the high plains of Texas, in Xinjiang, or in Uzbekistan. Cultivated cotton could survive only in frost-free climates. That's because cotton plants normally use the length of the day as a cue for when to bloom. They produce flowers and then seeds— and the fiber that surrounds them—only when the days grow short. (Some varieties also need cool temperatures.) In its native tropics, therefore, cotton may not flower until December or January, producing bolls in early spring. In places subject to frost, those plants wouldn't live long enough to reproduce.

That's why Mac Marston didn't quite trust his eyes when he looked at the sample under his microscope. Elizabeth Brite, a fellow UCLA graduate student in archaeology, had asked him to identify seeds she'd collected from Kara-tepe, a pre-Islamic site near the Aral Sea in northwestern Uzbekistan. Sometime in the fourth or fifth century CE, fire swept through a house there, carbonizing and preserving its contents, including a huge quantity of seeds that seemed to have been stored for future planting. By floating the seeds in a bucket of water and straining them through a sieve, Brite separated them from the surrounding dirt. She sealed the samples in vials about the size of film canisters and gave them to Marston. His job was to figure out what kinds of seeds they were.

"I was shocked when I put the first sample under the microscope and found that it was entirely cotton seed," recalls Marston, now at Boston University. "No. It's not cotton," he thought. "I'm making a mistake. It's something else. It looks kind of like cotton but it's got to be something else, because it shouldn't be there." Nobody expected to find cotton that far north—not at a site dating to no later than 500 CE. But the samples were excellent, the seeds were unmistakably cotton, and there were too many for them to be casual debris. People in Kara-tepe had been raising cotton.

Leaving aside the problem of frost, it made sense. Cotton needs full sun, hot weather, and not too much rain. So it was well adapted to the hot, arid region, with its salty soil and a river that swelled in the late spring and early summer, providing water for irrigation. Its life cycle complemented local food crops. And people in Kara-tepe could have gotten the seeds.

"This is an area that had clear trade contacts with India," says Marston. "So it's not like we found corn or something that would just be impossible" because it grew only on the other side of the world. But why would Indian farmers have identified and bred cotton that could grow once transplanted to Kara-tepe? Why would people in a frost-free region care about plants that weren't sensitive to day length?

Perhaps the change was driven by commercial competition. Suppose you're growing cotton in the Indus Valley, known as a source of cotton cloth as far back as Herodotus's writings in the fifth century BCE. If your cotton trees—and they were in fact trees—bloom earlier than your neighbor's, you can get to market earlier. You get paid sooner. Depending on how eager the buyers are, you may even command a higher price. The earlier the cotton harvest, the better off the farmer.

Over time, then, profit-seeking cultivators might favor early-blooming trees that weren't sensitive to day length. They'd replant, or perhaps sell, the seeds from those trees. Competition would push the flowering period earlier and earlier, until the harvest that once waited for winter came in the late summer and early fall. Farmers wouldn't have to know or care that the cotton was no longer sensitive to day length. They wouldn't have to consider frost. They'd just need to favor plants that would give them an earlier harvest. By doing so, they'd gradually develop breeds of cotton that could bloom even in places like Kara-tepe. In those northern climes, the frost would still kill the cotton plants—but only after the harvest. A new crop would need to be replanted in the spring. Instead of trees in orchards, cotton raised in colder regions became an annual row crop.[8]

Except for that last stage, we don't know what actually happened. But we do know that for cotton to grow in northern Uzbekistan, one way or another human beings first had to alter its nature. "People aren't going to bring it up there and start cultivating it unless this change—this biological, genetic change—to the crop has already happened," says Marston. "That being said, I don't think we actually discovered the first instance of this new, genetically modified crop." Like the linen cloth in the Nahal Hemar Cave, the cotton seeds at Kara-tepe are signs of a major innovation that had already become a well-established practice.

It would become even better established in the succeeding centuries, as the Islamic caliphate spread the cultivation of early-blooming cotton along with its new faith. Islam promised the faithful silks in heaven but forbade them to Muslim men in this world. Wearing cotton became a mark of devotion, and the demand for cotton grew with each new convert. "Plain white cotton (or linen in Egypt) signaled sincere Islam and marked its wearer as one who shared in the aesthetic of the conquering Arabs," writes historian Richard Bulliet.

In the wake of the Muslim conquest, he argues, cotton cultivation and trade led to the emergence of the Iranian plateau as "the most productive and culturally vigorous region of the Islamic caliphate." Beginning in the ninth century CE, Muslim entrepreneurs, mostly likely Arab transplants from Yemen, began creating new towns in arid areas such as the Qom region. They claimed land under an Islamic law giving ownership to anyone who brings "dead land" into cultivation. To irrigate their crops, they installed underground channels, or *qanats*. Although expensive to build, qanats could draw water year-round from the surrounding mountains and were well suited to growing cotton, which commanded a higher price than staple grains. "Unlike wheat and barley, which were normally grown as winter crops," writes Bulliet, "cotton was a summer crop that needed both a long, hot growing season and the steady irrigation that a qanat could supply."

The spread of cotton, much of it exported to Iraq, in turn fed the growth of Islam. The promise of financial rewards attracted workers to the new villages, where they adhered to the upstart faith. Converting gave Zoroastrian landowners less claim on the migrants' labors, making it harder to force people to return to their old homes. "In this fashion," Bulliet observes, "the cotton industry contributed to Islam's rapid spread in the rural districts close to key Arab governing and garrison centers." Within a century, the new villages had burgeoned into cities. The Muslim entrepreneurs, many of them religious scholars, grew extremely wealthy.

Versions of what happened in Iran occurred throughout the Muslim world. Islam fed the demand for cotton and Muslim cultivators increased its supply. "By the 10th century AD," write Brite and Marston, "cotton was found growing in nearly every region of the Muslim world, from Mesopotamia and Syria to Asia Minor, and from Egypt and the Maghreb to Spain."[9]

When Spaniards encountered cotton in the Americas, they knew exactly what they were looking at.

ᛗᛗᛗᛗ

From Mexico south to Ecuador, cotton was one of the treasures of the New World. Native peoples used finely woven cotton cloth for tribute, trade goods, and ceremonial objects. Cotton sails powered the oceangoing balsa rafts that traded along the Pacific Coast of Latin America. Cotton batting padded the cloth and leather armor of Aztec and Inca warriors. Cotton furnished the cords for the quipus on which the Incas kept records encoded in knots. When the Incas first faced the Spanish in battle, their cotton tents extended for three and a half miles. "So many tents were visible that it truly frightened us," wrote a Spanish chronicler. "We never thought that Indians could maintain such a magnificent estate nor have so many tents."[10]

Until the early nineteenth century, however, cotton cultivation in the Americas was largely confined to the tropics. Luxurious, long-fiber Sea Island cotton, a variety of *G. barbadense*, could grow in a few warmer areas on the US coast, but efforts to raise it in the rest of the South proved frustrating thanks to killing frosts. The two cotton varieties that would bloom before the winter frost were disease-prone, and their small bolls were hard to pick and clean. Planters longed for a cotton breed that would thrive in the fertile lands of the lower Mississippi valley, the early republic's southwestern frontier.[11]

In 1806, Walter Burling found it in Mexico City.

Burling was the kind of amoral adventurer who gives early capitalists a bad name. In 1786, while in his early twenties, he killed the father of his young nephew—whether his sister had married secretly is disputed—in a duel. Six days later, lured by the money to be made in human trafficking, he formed a partnership to enter the slave trade in what is now Haiti. When the island's enslaved population rebelled in 1791, initiating the Haitian Revolution, Burling was shot in the thigh and went back to Boston. In 1798, he embarked on the first American voyage to Japan, returning two years later with a cargo that included Japanese artifacts and a hold packed with Javanese coffee.

Burling married a Boston woman, then headed for the frontier, settling in Natchez, Mississippi, around 1803. Within a few years, he became the aide-de-camp to another amoral adventurer: General James Wilkinson, governor of the Louisiana Territory, partner with Aaron Burr in a plot to establish an independent country in the southwest, and secretly a spy for Spain. It was Wilkinson who sent Burling to Mexico City. His mission was to give the Spanish viceroy a letter from Wilkinson requesting a $122,000 payment for thwarting Burr's plot to invade Mexico and, while he was at it, to map possible invasion routes for the US government. Wilkinson was the kind of guy who worked every side, as long as he got paid.

Burling didn't collect the money; Spain apparently believed it had sufficiently compensated Wilkinson already. But he did spot a cotton variety that he thought might flourish in Mississippi and smuggled its seed back into the United States. In the surely apocryphal version of the story long taught to Mississippi schoolchildren, Burling asked the viceroy's permission to bring back the seed, only to be told that such exports were illegal, but "Mr. Burling could take as many *dolls* home with him as he chose; the dolls being understood *to be stuffed with cotton seed.*" Burling died in 1810, leaving no will and a pile of debts.[12] But his Mexican discovery changed history.

The new cotton variety indeed proved ideal for the Mississippi frontier. It ripened early, avoiding the frost. The bolls all appeared at about the same time, allowing an efficient harvest, and they were large and opened especially wide, which made them much easier to pick. "Because of this unusual quality," writes agricultural historian John Hebron Moore, "pickers could gather three to four times as much Mexican in a day as they could the common Georgia Green Seed cotton" previously in use. The ratio of fiber to seed was significantly better, yielding about a third more usable cotton after ginning. And the Mexican cotton was immune to a disease called the rot, which threatened to wipe out the area's cotton production. By the 1820s, farmers in the lower Mississippi valley had widely adopted the new breed.

They also improved it, both accidentally and deliberately. By carelessly allowing it to cross-pollinate with Georgia Green Seed, they inadvertently created a hybrid that preserved most of the advantages of the Mexican variety while eliminating its biggest drawback: the tendency of bolls to drop on the ground if not immediately harvested. Seed breeders then worked intentionally to refine the stock. By the early 1830s, a new Mexican-based hybrid called Petit Gulf dominated the Mississippi valley and was flourishing in the red clay soils farther east.

Burling's discovery, declares Moore, "improved the yield and quality of American cotton to such an extent that it deserves to rank alongside Eli Whitney's gin in the Old South's hall of fame." Patented in 1794, Whitney's invention—and a less acclaimed but more successful saw-based design by Hodgen Holmes a couple of years later—used a roller and brush to separate the cotton seed from the lint, mechanizing a previously laborious process and greatly increasing the potential supply of cotton.[13]

With good seed in hand, new gin technologies for processing cotton, and burgeoning demand from the mills of northern England, the "cotton fever"

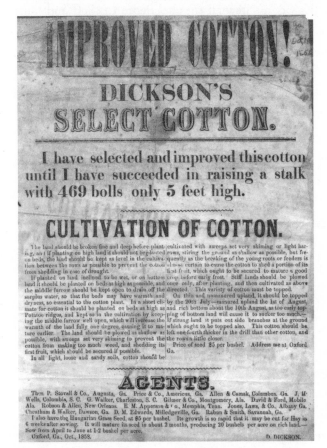

Cotton seed advertisement, 1858. Ads with the same text ran in many agricultural publications in the 1850s. (Duke University Library, Emergence of Advertising in America: 1850–1920 collection)

that had drawn pioneers like Burling to the frontier intensified. "Demand for U.S. cotton grew at better than 5 percent per year down to 1860, and the South emerged as nearly an ideal cotton-growing region in the pre-irrigation era," writes an economic historian. "It was said that American upland cotton could not be matched for 'uniting strength of fibre with smoothness and length of staple.'" There was big money to be made on the cotton frontier. From 1810 to 1850, Mississippi's population grew nearly fifteen-fold, from 40,352 to 606,526.[14]

Not all the Mississippi valley pioneers were ambitious planters with dreams of cotton wealth. Nearly half—a million people in the half century before emancipation—were enslaved workers forcibly uprooted from their families, friends, and familiar surroundings. This wrenching experience constituted a second exile, a replaying on American soil of the Middle Passage from Africa. Its victims likened the experience to theft and kidnapping. "They stole her back in Virginny and brung her to Mississippi and sold her to Marse Berry," former slave Jane Sutton recalled her grandmother saying.[15]

In some cases, the involuntary migrants were free citizens kidnapped by slave traders, as in the case of Solomon Northrup, whose memoir *Twelve Years a Slave* was the basis of the Oscar-winning 2013 movie by that name. More often, they were slaves whose eastern owners sold them to pay off debts or simply to profit from the demand for western labor. Slave traders crowded these unfortunate souls into ships bound for New Orleans or marched them hundreds of miles west, chaining the men together. Such slave coffles were a common sight on roads in the late summer and early fall, when the weather was favorable for the two-month trek.

Other enslaved migrants came west with their owners, often forced to leave behind spouses and children. "My dear daughter—I have for some time had hope of seeing you once more in this world, but now that hope is entirely gone forever," wrote Phebe Brownrigg to her free daughter Amy Nixon, shortly before her owner took her from North Carolina to Mississippi in 1835. One of the rare letters written by a western-bound slave on her own behalf, it concluded, "May we all meet around our Father's throne in heaven, never no more to depart."

Americans could have settled and cultivated the cotton frontier without slaves. After all, following the Civil War and emancipation, cotton production quickly rebounded and surpassed previous levels, with small farms supplying an increasing amount of the crop. But attracting voluntary migrants to the hardships of frontier life, and the region's hot, humid, and disease-ridden conditions, would have taken significantly longer. By forcing enslaved workers to move, cotton planters could quickly bring new land under cultivation.

"Planters and slave traders imported slaves at a higher rate than white pioneers migrated," notes a historian. "By 1835, Mississippi had a black majority." Fertile land and improved seed accelerated the expansion of slavery and made it more profitable. In a country where labor was the scarcest resource, cotton-growing pioneers had a workforce that couldn't quit and could even serve as collateral to finance their operations.[16]

In the popular imagination, the antebellum South is a technologically backward place, complacent and traditional—the antithesis of Yankee ingenuity. Even the cotton gin came from a New England inventor. In reality, the South nurtured its own scientific and technological ambitions, focused more on agriculture than on manufacturing. Holmes, whose saw gin surpassed Whitney's roller-based design, was from Savannah. Before Cyrus McCormick's mechanical reaper conquered the wheat fields of the Midwest, it was born on a Virginia plantation, with assistance from an enslaved man named Jo Anderson.[17] Slavery was inhumane, not incompatible with innovation.

Images of the antebellum South as technologically stagnant also confuse "technology" with machines, obscuring equally significant forms such as hybrid seeds. Unlike their Northern counterparts, Southern planters weren't primarily interested in labor-saving devices. They craved innovations that would get more out of their land and enslaved workforce. They rewarded entrepreneurs whose seeds promised higher yields.

"There has been evidently much improvement made in cotton within the past 20 or 30 years, and this has been done entirely by selection," wrote Martin W. Philips, a scientifically minded Mississippi planter, in 1847.[18] Thanks to improved plant breeds, from 1800 to 1860 the average amount of cotton picked per day per worker in the Southern states quadrupled from about twenty-five pounds to about one hundred pounds. (The best pickers could do much better, while others did worse.)

The demand for better cotton breeds was particularly concentrated in the newer states along the Mississippi, as was the innovation. "Most of the technologies were developed in the Mississippi Valley," write economic historians Alan Olmstead and Paul Rhode, who analyzed hundreds of plantation harvest records to track the effects of new seeds, "and were better suited for the geoclimatic conditions found there than for the conditions common to much of Georgia and the Carolinas, let alone conditions in India and Africa." As fields became more productive, Southern cotton cultivation steadily shifted west.[19]

Clever cotton breeding thus had enormous human and historical ramifications. Improved cotton encouraged the movement west, including the forced migration of enslaved workers. It further entrenched the economic role of slavery and deepened the divisions between the free North and the slave South that would eventually lead to the American Civil War. It increased supplies to British and New England mills, furthering the industrial takeoff that would raise global living standards to historically unprecedented levels. It gave US cotton producers an edge over farmers in India, the West Indies, and elsewhere.

Cotton breeders didn't have these geopolitical consequences in mind any more than they envisioned the blues and jazz, the novels of William Faulkner and Toni Morrison, or the jeans and T-shirts that came to symbolize youth and freedom in the late twentieth century. They were simply trying to grow more and better cotton. But textiles are never isolated from the rest of human life. For good and ill, they weave themselves into the fabric of civilization.

🀫🀫🀫🀫

Sericulture, the raising and harvesting of silkworms, is an ancient art. Silk proteins have been found in the soil under bodies in Chinese tombs eighty-five hundred years old. The proteins' locations suggest that the deceased were buried in the fabric, perhaps made from wild cocoons. Over time, Chinese cultivators turned wild caterpillars into the domesticated *Bombyx mori*, or mulberry silkworm, and harvested the filaments from its cocoons. The earliest silk fabrics discovered so far date back around fifty-five hundred years and appear to have been used to wrap a body before burying it in a coffin shaped like a silkworm pupa. By the Shang dynasty (1600–1050 BCE), sericulture was sufficiently well established to be the common subject of divinations and religious sacrifices.[20]

Over the millennia, as *Bombyx mori* was bred to human purposes, the insect came to depend on human protection. The adult moth is flightless—the better to keep it captive—and lacks the camouflage coloring that would allow it to survive in the wild. To produce silk, cultivators supply caterpillars with fresh mulberry leaves, raising the insects on trays protected from the weather. They give the developing silkworms sticks on which to build cocoons and then carefully observe the hibernation process. "From the day we swept the eggs onto the cartons," an old woman picking mulberry leaves told a Song dynasty traveler, "we have been tending to them as if they were newborn infants."[21]

Just before the moths emerge, the nurturing ends. Silk farmers harvest the cocoons and heat them to kill the insects so the moths can't break out and damage the silk. Only a few insects are allowed to emerge and reproduce.

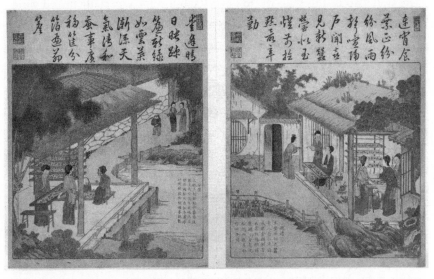

Tending silkworms, from Yu zhi Geng zhi quan tu (Pictures of Tilling and Weaving), *published in 1696 (Chinese Rare Book Collection, Library of Congress)*

Every stage in this process requires precision: just the right density of silk-worms and leaves, just the right temperatures, just the right timing. Incremental improvements can make a big difference.

During the Song dynasty (960–1279 CE), the demand for silk rose. To pay peace-keeping tributes to neighboring kingdoms, clothe an expanding army, and keep up court appearances, the government increased the taxes it required in silk yarn and cloth. Urban artisans, meanwhile, were buying ever more silk for weaving luxury fabrics to sell to the burgeoning bureaucracy. Like cotton planters in the American South, peasant farmers looked for ways to get more silk out of the same amount of land and labor. To this end, writes textile scholar Angela Yu-Yun Sheng, they "devised new techniques of production that looked simple in retrospect but in fact were ingenious. These new methods saved time and increased output."

Silk farmers figured out how to combine mulberry trees from two different regions of China, grafting the especially leafy mulberry species called Lu onto a sturdier trunk known as Jing. They also developed pruning methods that enhanced leaf production. These two improvements yielded a robust, year-round supply of silkworm food. With it, farmers could raise silkworms that reproduced several times a year, known as *polyvoltine* insects. Typically, there were two or three silkworm harvests, but some especially prized varieties could produce as many as eight generations in a single year.

As with cotton, an ideal crop of silkworms would mature at the same time but not go bad before it could be processed. So farmers came up with technological tricks to space out and coordinate the harvests. To control when silkworm eggs hatched, they learned to regulate their temperature. They spread eggs out onto thick paper cards and stacked ten or so cards together in a clay jar, which was immersed in cold water. They periodically took the cards out of the jar and allowed them to warm in the sun before immersing them again. Along with delaying when the eggs hatched, the process had a Darwinian effect. "As only strong silkworm eggs could survive both the cold and the wind," observes Sheng, "this method had the added advantage of weeding out the weak eggs."

Once the eggs hatched, farmers kept the silkworms well fed with mulberry leaves. To mature as quickly as possible, hastening the silk harvest, the insects needed a warm environment. Heating posed a technological dilemma, because the available fuels had distinct drawbacks. Wood smoke could damage the silkworms; burning manure wasn't harmful, but it didn't produce as much heat.

One solution was a portable stove in which wood could be burned outside to heat it up, then covered with ashes or manure and brought into the silkworm room. Another approach, preferred by large-scale breeders, was to dig a pit in the middle of the room, fill it with layers of dry wood and manure, and light the layers on fire about a week before the eggs hatched. The fire would steadily burn until a day or so before the silkworms emerged. At that point, cultivators would open the door just long enough to allow the smoke to escape, then shut it to keep the room warm as the silkworms hatched and matured. With these two methods, writes Sheng, "peasants in the Song shortened the caterpillar's growing time during the second morphological stage," when it repeatedly molts before spinning its cocoon, "from 34 or 35 days down to 29 or 30 days, or even to as few as 25 days."

Silkworm cultivators also discovered that once the cocoons were ready for harvest, they could be preserved for an additional week by sprinkling them with salt. This invention allowed the painstaking process of reeling the silk off the cocoon to be performed over time, permitting the same number of people to produce more silk per harvest. As an added benefit, the salt improved the silk's quality.

Alone, none of these innovations was momentous, but together they allowed farmers to produce significantly more silk with the same amount of land and labor. That productivity boost enabled them to bear the burden of heavier taxes while still taking advantage of new commercial markets. Some peasants abandoned subsistence farming altogether to concentrate on textile production.[22] As with cotton in the Old South, the story of silk in Song China demonstrates that technological innovations need not involve machines.

ⴕⴕⴕⴕ

Nature includes not only the animals and plants from which humans can obtain fibers but also the enemies that threaten to destroy them—and not all such menaces are as easily identifiable as the Cotton South's notorious boll weevil. The microbiology that revolutionized human understanding of infectious disease, saving the lives of millions, began with a quest to rescue silk production.

Around the time that Walter Burling was smuggling Mexican cotton seed into Mississippi, a curious Italian began experiments to figure out why silkworms were dying in droves. The twin son of peasant farmers, Agostino Bassi trained as a lawyer and held various bureaucratic posts in Lodi, a town about twenty miles south of Milan. His real passion, however, was for science and medicine. Using the family farm as his laboratory, Bassi conducted experiments and published treatises on raising sheep, cultivating potatoes, aging cheese, and making wine. His most important—and time-consuming—research was on silkworms.

In late 1807, the thirty-four-year-old Bassi embarked on what turned out to be thirty years of experiments aimed at identifying and countering the cause of a disease variously known as *mal del segno*, muscardine, or, in a nod to the white powder covering the caterpillars it killed, *calco, calcino,* or *calcinaccio.* Silkworms would stop eating, become limp, and die. Their corpses would then grow stiff, brittle, and coated in white. Breeders believed that the disease must be caused by something in the insects' environment, and Bassi set out to figure out what it was.

His first eight years of experiments proved frustrating and apparently futile. He later wrote:

> I used many different methods, subjecting the insects to the cruelest treatments, employing numerous poisons—mineral, plant, and animal. I tried simple substances and compounds; irritating, corrosive, and caustic; acidic and alkaline; soils and metals; solids, liquids, and gases—all the most harmful substances known to be fatal to animal organisms. Everything failed. There was no chemical compound or pest that would generate this terrible disease in the silkworms.

By 1816, Bassi was deeply discouraged. He had expended enormous effort and nearly all his money on fruitless studies. He was losing his eyesight.

"Oppressed by a great melancholy," he abandoned his research. But a year later, he rallied, resolving to "defy misfortune, turning to interrogate nature in new ways with the firm resolution of never abandoning her until she responded sincerely to my questions."

A major clue came when Bassi observed that silkworms raised in the same conditions and fed the same food but housed in adjacent rooms had different outcomes. The disease would sweep through one room while its neighbor suffered little or no damage. The difference, he concluded, was that "there was no *calcino* germ, or very few, in one room and large numbers in the other. The *mal del segno* or muscardine is never born spontaneously" in reaction to a toxin, as everyone had previously believed.

After more experiments, Bassi realized that living insects wouldn't infect one another. Rather, the disease was carried by the white coating that

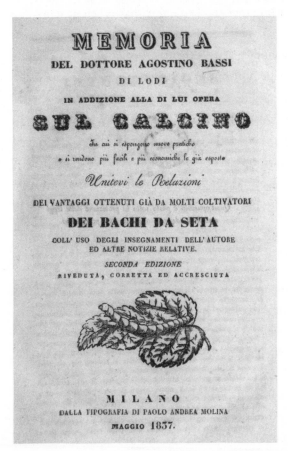

The germ theory of disease began with Agostino Bassi's investigation of the mysterious calcino ailment killing silkworms. (Wellcome Collection)

appeared on the corpses. Introduced into the body of a living insect, whether caterpillar, pupa, or moth, the powder would multiply inside, feeding on the insect's body until it killed it. Only then would it spread. "While it needs the life of the invaded individual to develop, grow, and render itself able to reproduce," wrote Bassi, "it doesn't produce its fruits or seeds, or at least they don't mature or they aren't fertilized, until it has extinguished the animal that received and fed it. . . . Only the corpse has the ability to contaminate." He concluded that the invader was a fungus and the white substance its spores.

By placing a dead insect in a warm, humid environment, Bassi found he could cultivate the fungus enough to detect hints of stems with the naked eye. Under a simple microscope, he could see the curves that marked the invader as a living organism rather than a crystal. With the powerful new compound microscope invented by Giovanni Battista Amici in 1824, Bassi wrote, one could see "all its minute branches and perhaps even its reproductive apparatus."

Having determined the culprit, Bassi experimented with ways of killing the fungi without harming the silkworms, identifying several effective disinfectants. To stem the pestilence, he advised sanitary measures that included treating all silkworm eggs with disinfecting solutions; boiling instruments; disinfecting trays, tables, and workers' clothing; and requiring everyone tending the silkworms to wash their hands with disinfectants.

As these hospital-style measures suggest, Bassi's discovery was a breakthrough with implications beyond sericulture. His research anticipated the more famous work of Louis Pasteur and Robert Koch in developing the germ theory of disease. The provincial lawyer was a scientist ahead of his time.

"For the first time, a human being formulated the parasite theory of disease," declares a journal article marking the bicentennial of Bassi's birth. The standard reference *Disinfection, Sterilization, and Preservation* calls Bassi's experiments "the first clear demonstration of the microbial origin of disease in animal life," noting that he extended his work to "a theory of contagion originated from living parasites in infected wounds, in gangrene, cholera, syphilis, plague, typhus, and the like. He suggested the use of germicides, mentioning alcohol, acids, alkalies [*sic*], chlorine, and sulfur."[23]

Nine years after Bassi's death in 1856, Pasteur, who was generously funded and much more adept at public relations than his Italian predecessor, undertook a similar scientific challenge. The French government engaged the famous man to investigate a new and even more devastating silkworm disease known as *pébrine*. When he started, Pasteur knew nothing about silkworms and in fact hadn't worked on any animal diseases. His previous research had been on fermentation and yeast. But he was supremely confident and a quick study. Among the resources he had at his disposal were French translations of Bassi's work.

Over five years of experiments, Pasteur developed a way to separate infected silkworm eggs from those that would produce caterpillars free of pébrine. He also identified another, sometimes overlapping, disease called flacherie, along with measures to keep it from spreading. The silkworm experiments introduced him to animal biology, altering the course of his scientific career. "The caterpillar of Alès led Pasteur from microbiology to veterinary science and to medicine," writes his biographer Patrice Debré, himself an immunologist. The path to Pasteur's vaccines against anthrax and rabies, and eventually to the great triumphs of public health that dramatically extended life expectancies, began with silk.[24]

𝍌𝍌𝍌𝍌

Pasteur did not cure pébrine. He merely found a way to identify and destroy infected eggs, ameliorating the plague but coming nowhere close to ending it. By the early 1860s, French silk production was a fifth of what it had been a decade earlier. Italian production had fallen by half.

For uninfected eggs, the European silk industry increasingly turned to Asia, particularly the newly open country of Japan. In European markets, Japanese silkworm eggs sold for ten times the price French eggs had once commanded. As a diplomatic gift, in 1864 the Tokugawa shogunate presented Napoleon III with fifteen thousand silkworm egg cards. Although China remained the primary exporter of raw silk, Japan surpassed its neighbor as Europe's most important source of eggs.[25]

Like its European counterpart, Japanese sericulture had its roots in China and had grown greatly beginning in the seventeenth century. As the shogunate restricted Chinese imports, expanding the domestic market, silk farmers began to specialize in either breeding eggs or raising silkworms. Through experimentation and meticulous attention, they gradually improved techniques, boosting quality and production. To feed growing silkworms, for instance, Japanese sericulturalists followed the Chinese practice of cutting up mulberry leaves. But they didn't stop there. They passed the leaves through increasingly fine sieves, reserving the smallest pieces for the youngest worms, giving larger bits to older insects, and eliminating any debris.

"The same minute attention was paid to virtually every other aspect of the silkworms' welfare," writes historian Tessa Morris-Suzuki. "Silkworm trays were frequently moved from one part of the house to another to protect them from heat or cold, and the amount of food given to the worms was varied according to changes of temperature. Trays and utensils were regularly washed and aired in the sun, and strict rules governed the personal cleanliness of silk workers."

In the early nineteenth century, a silk farmer named Nakamura Zen'emon began making his own thermometers, modeled on Dutch imports, and using them to conduct experiments. He found that some stages, such as egg production, required slightly higher temperatures, whereas others needed lower ones. In 1849, Nakamura published an illustrated manual for silk farmers, promulgating his results throughout Japan.

Japanese silkworms weren't immune to pébrine, muscardine, and other diseases. (Pasteur identified pébrine in some of the eggs given to Napoleon III.) But good practices made disease less likely to spread. Silk workers used leaves only from healthy mulberry trees, didn't crowd as many silkworms on trays, and removed any sick-looking caterpillars. They frequently washed their hands and changed their clothes—practices that Bassi would have applauded.

To maintain silkworm quality, Japanese sericulturalists bought eggs from specialists rather than relying on their own moths. Egg breeders grew wealthy by developing new hybrid insects, aiming to improve the quality and quantity of silk and to produce specific characteristics for specific uses. The combined result of better silkworm breeds and careful rearing techniques was a dramatic leap in productivity.

At the beginning of the nineteenth century, it took forty days to raise silkworms from hatching to cocoons, compared to fifty a century earlier. The amount of silk per cocoon rose by more than a third—and jumped another 40 percent in the first half of the nineteenth century. By the 1840s, Japan's sericulture practices had captured European attention, with the French translation of an illustrated sericulture manual published in 1848. The book, Morris-Suzuki observes, "became not only Japan's first technology export to the West but also one of the first Japanese works of any sort to be translated into a European language."

When US commodore Matthew Perry and his famous "black ships" arrived in Edo harbor in 1854, forcing Japan to open trade with the United States and, eventually, with other Western nations, Japan's sericulturalists were ready for world markets. The preceding two centuries had fostered a thriving industry with valuable products to export. Raw silk and silkworm eggs brought in money that could be invested in the railroads and factories Japan lacked.

Just as important, sericulturalists had created a Japanese culture primed to make the best of—and improve upon—foreign knowledge. "What is important about all this is not simply that Japanese raw silk production expanded and improved in quality during the Tokugawa period," writes Morris-Suzuki, "but that many silk farmers developed attitudes that recognized the need for experimentation, technological change, and even the incorporation of Western ideas like the thermometer into the production process."[26]

With the Meiji Restoration of 1868, Japan embarked on an official policy of modernization, vowing that "knowledge shall be sought throughout the world." A Japanese team spent a month studying at a new sericulture institute in northern Italy, returning with tools including state-of-the-art microscopes and hygrometers for measuring humidity. In 1872, the government sponsored the first reeling factory, importing French machinery; private ventures followed. By the mid-1890s, hand reeling accounted for less than half of the country's raw silk production.

The regional silk variations that had suited Japan's Tokugawa market, where fabric characteristics reflected fine gradations of fashion, identity, and status, proved a problem for industrial production. Japanese silk reeling factories "had long maintained that the production of so many different kinds of cocoons was the greatest single cause of the uneven quality of raw silk," writes an economic historian.

That changed in the 1910s, when Japanese scientists combined the country's long tradition of silkworm breeding with Mendelian genetics to develop a highly productive silkworm hybrid. Its cocoons were so superior, especially for machine reeling, that the new silkworm type swept the country. By establishing a de facto standard, the hybrid made Japanese silk more consistent. At the same time, newly precise temperature controls during reeling led to higher-quality thread.

Japanese raw silk provided the perfect supply for US silk mills. Established by English immigrants in the mid-Atlantic states of New Jersey (where Paterson became known as "Silk City"), New York, and Pennsylvania, US silk production differed from the European version. It relied on fast power looms to turn out large quantities of inexpensive standardized fabric—a democratized luxury for a continental market. Unlike the handloom weavers of France and Italy, American factories had no domestic sericulture to draw on, and Chinese silk was too irregular for high-speed, automated looms. The new Japanese variety was perfect. The US and Japanese silk businesses grew up together, each depending on the other. By the early twentieth century, these upstarts dominated world markets.[27]

☒☒☒☒

In 2009, almost exactly two hundred years after Bassi set out to discover what was killing Italian silkworms, three young Bay Area scientists started a company to reverse the relationship between microorganisms and silk and, in the process, give humans greater control over the fiber's properties. Rather than protecting silkworms from tiny predators, Bolt Threads turns microorganisms into silk-producing machines. Funded by some of Silicon Valley's leading venture capital firms, the company bioengineers yeast so that instead of alcohol the cells excrete silk proteins. The fermentation research with which Pasteur began his career has led back to silk in a way no nineteenth-century scientist would have anticipated.

As we tour Bolt's labs, David Breslauer, the company's chief scientific officer, takes a one-pound jar out of a cabinet and scoops out some off-white protein powder. The stuff looks ready for a smoothie. But this isn't a health food. Composed of the protein found in super-strong dragline spider silk, it's the main ingredient in the first entirely new fiber in decades. This one jar, says chief commercial officer Sue Levin, "is more silk protein powder than has ever been in one place at one time." To turn the powder into yarn, Bolt dissolves it into a molasses-like concoction that it extrudes and wet spins into fine, lustrous fibers that can be knitted or woven into cloth.[28]

Bolt defines itself as a biology-based materials company; it also makes a leather substitute called Mylo from *mycelium*, the cells that make up mushrooms. It casts what it calls Microsilk as the scaffolding for something much larger than spider's silk: the first entirely novel class of textiles since 1935, when DuPont chemist Wallace Carothers cooked up nylon and launched the polymer revolution. Bolt calls its textile products "protein-polymer microfibers."

Protein fibers aren't a completely new idea. Inspired by the success of rayon from regenerated wood pulp, scientists in the 1930s turned their attention to proteins. Henry Ford sponsored research on soybean-derived fibers, hoping to find a substitute for wool in automobile upholstery. Imperial Chemical Industries in Britain developed Ardil from peanuts. Others experimented with egg whites, zein protein from maize, and feathers.

The most successful new protein fiber was Lanital, an Italian invention derived from skim milk. Subsidized by the Fascist government to encourage national self-sufficiency, SNIA Viscosa, a leading rayon producer, turned out ten million pounds of Lanital in 1937. An American version, its producers boasted, was "man's first successful effort to create a fibre which could stand beside such natural protein fibres as wool, mohair, alpaca, camel's hair, and fur." Soft, warm, and shrink-resistant, milk-based materials sounded like good substitutes for wool. But they had distinct drawbacks. People thought they smelled like cheese or spoiled milk when wet. And they didn't hold up well. An Italian couturier remembered how her sister called Lanital the "mozzarella fabric, because when you ironed it threads came up like tiny threads of cheese." After the war, people went back to wool or bought synthetics such as polyester, nylon, and acrylic.[29]

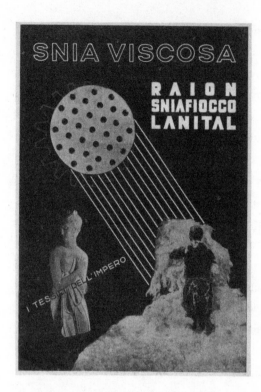

In the 1930s, exciting new fibers included rayon and Lanital, made from skim milk and subsidized by Italy's Fascist government to reduce wool imports. (Sniafiocco was the staple form of rayon.) (Author's collection)

It's those petroleum-derived polymers Bolt aims to replace. They changed the world with no more than a hundred possible combinations, observes CEO Dan Widmaier. Protein-based polymers offer many more possibilities. "*Every* functionality in every living organism on the planet is based on making protein polymers," he says. Molecular biologists understand how a DNA sequence becomes a structure, he says, so "if we can make dragline spider silk, the same way you make that protein and do it on a large scale, you can make pretty much any structural protein. When I run the math on how many that is, it's like 10^{106} polymers."

By selecting the right amino acid sequence, Bolt could imbue fibers with specific properties. The company envisions a nearly endless array of new materials, each tuned to specific needs: stretch, strength, fineness, UV resistance, breathability, water resistance, you name it. Breathable workout shirts that never stink. White sofa cushions that repel red wine. Hospital sheets that kill infectious microbes. Fabrics so flexible they literally feel like a second skin. Sweater material as soft as cashmere but without the microscopic scales that make supersensitive skin like mine itch—or the herds of goats whose appetites are turning Mongolia into a dust bowl. All of it produced with environmentally benign ingredients and able to break down naturally when thrown away.

If the endeavor succeeds, silk won't need to come from insects. It will be brewed in giant fermentation vats like beer. And the process has implications beyond silk. Wool, too, is a protein polymer. So is cashmere. So are countless other fibers we can barely imagine. And, like petrochemical polymers, proteins can form solids or gels as well. Bolt once experimented with solid silk buttons just to prove it could theoretically make an all-silk garment.

It's an intoxicating vision. But so far, the only Microsilk products are one-off stunts. In 2017, Bolt sold small runs of silk neckties and silk-wool blend caps. Excited about the prospect of vegan silk, fashion designer Stella McCartney featured Bolt's fiber in a few runway pieces and a gleaming yellow dress for a Museum of Modern Art exhibit. Two years later, she again used Microsilk, blended with the cellulose-based fiber Tencel, in a showcase tennis dress. British lingerie designers Strumpet & Pink crocheted a wispy pair of knickers for a show at the Rhode Island School of Design. And Bolt designers created eyeglass frames of solid silk protein, with a lanyard of Microsilk and a case of Mylo, to display at a biofabrication conference. "It was a little capsule of all these cool biomaterials coming together," says Widmaier.

But that's it: cool proofs of concept, not fabric in quantities designers can use for products to sell. Microsilk is not available in stores, nor is it likely to be for some time.

By 2019, four years after my first visit, the company had developed a supply chain that it says could produce tons of the stuff. But it didn't ramp up production, turning its attention instead to Mylo. Potential customers are more interested in a leather substitute than a protein microfiber. When you follow the money, says Widmaier, "you find accessories made out of leather and vegan leather." Bolt is a business, not a charity. Leather may be a much smaller market than textiles, but it offers higher profit margins and less competition.

Every new fiber idea eventually confronts the fundamental truth about textiles: Ancient and pervasive, they embody the experiments of countless

Detail of the Stella McCartney dress made from Bolt Threads' bioengineered Microsilk and featured in an exhibit at the Museum of Modern Art. (Bolt Threads)

generations. Human beings have been improving fibers for tens of thousands of years. Even synthetics have had eighty years of intense refinement. Only the best materials can survive the competition. Many, from maguey and nettles to Lanital and Ardil, have essentially vanished. Former mainstays, including wool and hemp, today occupy specialty niches.

Tapping scientific knowledge and technological tools unavailable until recently, Bolt believes it can beat the odds. It's betting on two factors: growing environmental concerns and all those functional possibilities. By fine-tuning proteins, its scientists can replace thousands of years of fiber breeding in a matter of days. "We've found that we can generate ideas and demos faster than we could ever really get them up to scale," says Widmaier. The trick, then, is to identify the formulations with the greatest commercial potential. Neither natural nor synthetic, bioengineered protein-polymer fibers could be the next stage in the process that began with selective breeding.[30]

Chapter Two

THREAD

When Adam delved and Eve span, who was then the gentleman?
—English saying

ON THE GROUND FLOOR OF Amsterdam's Rijksmuseum, two stories below
the Rembrandts and Vermeers, hangs a pair of sixteenth-century paintings that foreshadow the wealth that would fund Holland's artistic efflores-
cence. The twin portraits are of a young couple, believed to be Pieter Bicker
and his wife, Anna Codde. Painted in 1529 by Maarten van Heemskerck,
they are among the earliest portraits of Dutch citizens.

Depicted at a stylish three-quarters angle rather than in old-fashioned
profile, the two subjects are clearly specific individuals, not generic types.
Anna, a pale blonde with dreamy eyes and the slightest hint of a forehead
crease, complements her dark-haired husband, with his alert expression and
chiseled cheekbones.

They are Anna and Pieter, real historical people. Posing with the tools of
their trades, however, they could also serve as allegories. She sits at a spinning wheel, her left hand drawing out the yarn as it stretches toward the
spindle from a mass of fiber. He holds a ledger in his left hand and counts

Portraits of a couple, possibly Anna Codde and Pieter Gerritsz Bicker, by Maarten van Heemskerck, 1529 (Rijksmuseum)

coins with his right. Their hand positions mirror each other: One pinches the handle of a wheel; the other, a coin. One grasps thread between her thumb and three fingers; the other holds his account book the same way. Here are the endeavors essential to Dutch prosperity: Industry and Commerce personified.[1]

Today we associate *industry* with smokestacks. But they became its emblem only in the nineteenth century. From the Renaissance on, industry's visual representation was a woman spinning thread: diligent, productive, and absolutely essential.

Today's critics tend to emphasize the implied domesticity and subordination of the era's images of spinning women. "While Pieter Bicker is characterized as a bold and shrewd businessman, his wife is depicted spinning, again symbolic of the virtuous housewife," observes an art historian.[2] This view casts the woman spinning as passive, economically dependent, and culturally inferior—a contrast to the independent, public male merchant. It imagines the counting house as a real and significant enterprise while treating the spinning wheel as mere iconography, symbolizing the "virtuous housewife" the way keys might designate Saint Peter.

In fact, van Heemskerck's spinning wheel is as realistic—and economically indispensable—as his accounting ledger. "No picture could show more clearly the spinning of long fibres with one hand," writes Patricia Baines in her authoritative 1977 study *Spinning Wheels, Spinners and Spinning*. "They are drafted by the index finger and thumb, and a tension is kept on the thread between the third finger and the orifice," the opening that lets newly created yarn wind onto the bobbin. "One can see how the thumb is perfectly poised to roll along the index finger, and how the wrist is turned as the thread is smoothed before a few more fibres are selected."[3] Anna isn't just posing. She knows how to spin.

Dismissing spinning as a symbol of domestic submission rather than productive industry misses the reason *why* from antiquity onward it was honored as a sign of feminine virtue—or why, for that matter, the Industrial Revolution began with spinning machines. Only because thread has been plentiful for two hundred years does producing it seem like something other than the epitome of fruitful labor. Throughout most of human history, producing enough yarn to make cloth was so time-consuming that this essential raw material was always in short supply. The quest for thread prompted some of the world's most important mechanical innovations, leading ultimately to the Great Enrichment that lifted worldwide living standards. The story of thread illustrates how, despite immediate dislocations, labor-saving technologies can create abundance and free people's time for more economically valuable and personally satisfying purposes.

꙰꙰꙰꙰

A *spindle whorl* doesn't look like much. It's a small cone, disk, or sphere made of a hard material such as stone, clay, or wood, with a hole through the center. Museums often own thousands but display only a handful to the public. Even these special few, singled out for their decorative paint or engraving, are easily overlooked in favor of the showier vases, bowls, and statuettes nearby. "Spindle whorls are not the most spectacular objects found by archaeologists," concedes a researcher.[4] They are, however, among the earliest and most important human technologies—a simple machine as essential

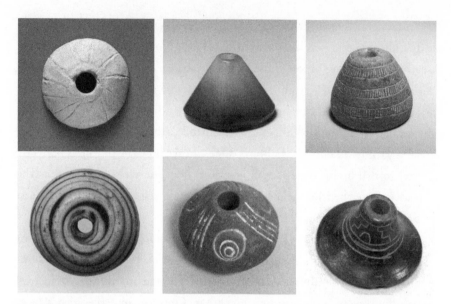

Spindle whorls, clockwise from top left: Sumerian, ceramic, 2900–2600 BCE; Minoan, agate, 2450–2200 BCE; Cypriot, terracotta, 1900–1725 BCE; Roman, glass, first to second century CE; Peruvian, possibly north coast, ceramic with pigments, 1–500 CE; Mexican, ceramic, tenth to early sixteenth century CE (Metropolitan Museum of Art)

as agriculture in going from small amounts of string to the large quantities of yarn needed for making cloth.

"The spindle was the first wheel," Elizabeth Barber tells me, gesturing to demonstrate. "It wasn't yet load bearing, but the principle of rotation is there." A linguist by training and weaver by avocation, Barber started noticing footnotes about textiles scattered through the archaeological literature in the 1970s. She thought she'd spend nine months pulling together what was known. Her little project turned into a decades-long exploration that helped to turn textile archaeology into a full-blown field. Textile production, Barber writes, "is older than pottery or metallurgy and perhaps even than agriculture and stock-breeding."[5] And textile production depends on spinning.

Leaving aside silk, which we'll get to later, even the best plant and animal fibers are short, weak, and disorderly. Flax fibers can reach a foot or two, but six inches is as long as a strand of wool will grow. Cotton typically runs only an eighth of an inch, with the most luxurious varieties extending no more than two and a half inches. Drawing out these short fibers, known generically as *staple*, and twisting them together—spinning, in other words—produces strong yarn, as individual fibers wind together in a helix, creating friction as they rub against each other. "The harder you pull lengthwise, the harder the fibers press against each other crosswise," explains a biomechanics researcher.[6] Spinning also extends the length of staple fibers, producing thread that can stretch for miles if needed—as it usually is.

Spindle whorls are the durable components of a two-part mechanism that, with slight variations, many different people invented in many different places, from China to Mali, the Andes to the Aegean. A stick goes through the hole, with the spindle whorl near one end. To make thread, a spinner stretches out a bit of fiber still attached to a clean mass of wool, flax, or cotton that has been brushed so that the strands run more or less in the same direction. She ties some twisted fiber around the rod, then drops the apparatus while setting it spinning. The whorl adds weight and increases the spindle's angular momentum, allowing the spinner to maintain the rotation while continuing to add new fiber as gravity stretches the thread downward. When the string grows too long for her to hold it off the ground, the spinner winds the new yarn around the stick, preserving its twist. These

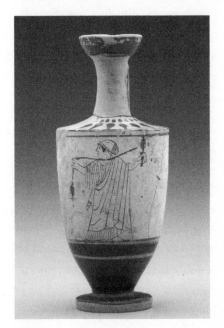

three steps—*drafting, twisting,* and *winding*—together constitute the process of spinning yarn.

In the hands of an experienced spinner using well-prepared fiber, it looks effortless. The yarn seems to grow on its own. But spinning is quite tricky. You have to keep exactly the right tension on the continuous supply of new fiber—enough to make the thread fine and even, but not so much that it breaks off—while maintaining a steady amount of spin. In a six-hour drop-spinning workshop, with generous hands-on assistance, I managed to produce about ten yards of uneven, two-ply wool yarn. If that

Greek vase circa 460 BCE (Yale University Art Gallery)

sounds impressive, picture a ball about an inch in diameter. Fine for knots. No good for cloth.

Once you get the feel for it, however, spinning becomes second nature. Hobbyists call it a stress reliever. "When you first begin," says spinner Sheila Bosworth, "no one would ever believe it is calming and relaxing, but once you know what you are doing, the rhythm is very meditative." Bosworth takes her *drop spindle* with her everywhere, spinning as she waits in line, sits in restaurants, or rides in the car.[7]

In this, she mimics the countless generations who spun out of necessity. Using a drop spindle, preindustrial spinners could work while minding children or tending flocks, gossiping or shopping, or waiting for a pot to boil. They could spin indoors or out, in company or alone, in close quarters or open space.

Although not quite as portable, spinning wheels were often light enough to take outside in good weather. Renaissance Florence banned spinners from crowding its public benches. Traveling through Suffolk and Norfolk in the late seventeenth century, diarist Celia Fiennes observed women "at their wheeles out in the streets and Lanes as one passes." In northern Europe, spinners brought their tools to communal "spinning bees," sharing heat, light, and sometimes-raucous companionship on long winter nights. "Alone she does not earn her lighting costs" was how a German farmer's wife in 1734 justified defying a local ban on such gatherings. For others, the attraction lay in the fellowship, including the young men who dropped by to flirt with the spinners.[8]

Unlike other steps in textile production, spinning has almost always been done exclusively by women. "That woman alone is good who works all the time with the *charkha*," or spinning wheel, wrote the Indian historian and poet Abdul Malik Isami in 1350.[9] The English word *distaff* is both a tool to hold fiber while spinning and an adjective denoting something related to women, most often the mother's or wife's side of a family. *Spinster* means both someone who spins and an unmarried woman.

On ancient Greek pottery, spinning appears as both the signature activity of the good housewife and something prostitutes do between clients. "In the same manner that sex was the trade of the prostitutes, so too was the making of textiles," writes an art historian.[10] A similar contrast appears in sixteenth- and seventeenth-century European art. In expensive paintings like the portrait of Anna Codde, spinning exemplifies domestic industry and virtue. In popular prints, it's often sexually charged. A Dutch engraving from 1624 shows a young woman with a large, fiber-filled distaff—a pole whose contents give it an especially phallic shape—under her right arm. Her left hand fondles the fibers, holding the distaff's bulging end unrealistically close to her face. Rather than drawing out fibers to spin, she seems on the verge of kissing them. The accompanying text transforms spinning into an extended sexual metaphor.

> I am stretched long, white—so you see—and fragile. At the uppermost am I the head, slightly big. My mistress wishes me steady, often has me in her lap; or instead, she lays me nearby her side. She holds me many times—yes, daily, may I say, with her hands. She pulls her knees up, and in a rough place, now she sticks my top. Now she pulls it out again. Now she goes to place it again.[11]

Virtuous or sexy, take your pick. Whatever a man wanted in a woman— or a woman aspired to be—spinning could represent it. Whether designed to valorize or titillate, these popular images reflected an everyday reality. Most preindustrial women spent their lives spinning. Unlike weaving, dyeing, or raising sheep, it was less a specific occupation than a universal life skill like cooking or cleaning. A poor woman might do it for money, just as she might hire out as a maid, but she had that option because she'd learned from childhood how to spin—and because there was always demand for more thread. Always.

At only four years old, an Aztec girl was introduced to spinning tools. By age six, she was making her first yarn. If she slacked off or spun poorly, her mother punished her by pricking her wrists with thorns, beating her with a stick, or forcing her to inhale chili smoke. The severity of the punishments reflected the importance of mastering the craft.

Along with the bast fibers they spun for their own families' use, Aztec women had to produce enormous amounts of cotton yarn to meet the tribute demands of the empire's rulers. Every six months, for instance, the five towns in the conquered province of Tzicoac paid a tax of sixteen thousand white cloaks bordered in patterns of red, blue, green, and yellow, along with similarly astonishing amounts of underclothes, oversized white cloaks, and women's clothing. "Only successive generations of frenetically productive spinners could meet the enormous demands for clothing tribute," observes a textile historian.[12]

Whether Aztec mothers, orphans in the Florentine Ospedale degli Innocenti, widows in South India, or country wives in Georgian England, women through the centuries spent their lives spinning, especially after water wheels freed up time previously devoted to grinding grain.[13] Preindustrial women spun constantly because cloth, whether for taxes, sale, or household use, required a huge amount of thread. Today we have the luxury of taking that thread for granted.

Consider jeans. Today's average Mexican, descended from the women who once spun cotton for imperial tribute, owns seven pairs, the average American, six, and the average Chinese or Indian, three. Weaving the denim in a single pair requires more than six miles, or nearly ten kilometers, of cotton yarn.[14] Working eight hours a day, a spinner using a traditional Indian charkha would take about twelve and a half days to produce that much thread—not including the time to clean and comb the fibers for spinning. If all that cotton had to be spun by hand, even at poverty wages, jeans would be luxuries.[15]

This example actually underplays just how much work spinning once required. For starters, jeans don't demand all that much yarn, thanks to their relatively coarse weave of about a hundred threads to the inch, and the small amount of fabric they require. Other everyday essentials need much more. Consider a twin bedsheet with a modest thread count of 250. Weaving it requires about twenty-nine miles of thread—enough to extend from downtown San Francisco to the Stanford University campus or from Kyoto to Osaka. A queen-size sheet takes about thirty-seven miles, which would stretch from the Washington Monument to Baltimore or from the Eiffel Tower to Fontainebleau.[16]

In addition, the charkha, which uses a larger wheel to turn the spindle many times with a single rotation, is one of the fastest ways to hand-spin

	AMOUNT OF YARN REQUIRED	CHARKHA, COTTON (100 M/H)	SPINNING WHEEL, WOOL, MEDIUM (91 M/H)	ANDEAN, WOOL (90 M/H)	VIKING, WOOL, COARSE (50 M/H)	ROMAN, WOOL (44 M/H)	EWE, COTTON (37 M/H)	BRONZE AGE, WOOL, FINE (34 M/H)
JEANS/ TROUSERS	6 miles = 10 km	100 hours = 13 days	110 hours = 14 days	111 hours = 14 days	200 hours = 25 days	227 hours = 28 days	270 hours = 34 days	294 hours = 37 days
TWIN SHEET	29 miles = 47 km	470 hours = 59 days	516 hours = 65 days	522 hours = 65 days	940 hours = 117 days	1068 hours = 134 days	1270 hours = 159 days	1382 hours = 206 days
QUEEN SHEET	37 miles = 60 km	600 hours = 75 days	659 hours = 82 days	667 hours = 83 days	1200 hours = 150 days	1364 hours = 171 days	1621 hours = 203 days	1765 hours = 221 days
TOGA	25 miles = 40 km	400 hours = 50 days	440 hours = 55 days	444 hours = 56 days	800 hours = 100 days	909 hours = 114 days	1081 hours = 135 days	1176 hours = 147 days
SAIL	60 miles = 154 km	1540 hours = 193 days	1692 hours = 211 days	1711 hours = 214 days	3088 hours = 385 days	3500 hours = 438 days	3621 hours = 453 days	4529 hours = 566 days

Note: *This chart compares the amount of time it takes to spin a certain length of yarn with a given technique. It is designed to give a general picture and reasonable estimates, not precise equivalencies, and does not account for the fiber actually used in the item. Cotton is typically harder to spin than wool. Calculations assume eight-hour workdays.*

yarn. Spinning enough cotton to make a traditional Ewe woman's cloth, roughly the equivalent of enough fabric for a pair of jeans, takes a West African spinner about seventeen days. On the eve of the Industrial Revolution, in the eighteenth century, Yorkshire wool spinners using state-of-the-art treadle spinning wheels would have taken fourteen days to make that much yarn, and wool is easier to spin than cotton.[17]

Andean spinners, who work with sheep's wool and alpaca fiber on drop spindles, spin about ninety-eight yards an hour. This translates to about a week to produce enough for a square yard of fabric, which was the off-the-cuff estimate of the Peruvians who taught my drop-spinning workshop.[18] Keep it up and in a couple of weeks you'll have enough to weave the cloth for a pair of pants. Not surprisingly, today's Andean spinners buy factory-made trousers, reserving their handmade yarn for less quotidian uses.

Even this time-consuming process is faster than some ancient methods. Experienced spinners using re-creations of Bronze Age drop spindles can generate between thirty-four and fifty meters of wool yarn an hour, depending on how finely they spin. (Finer thread, produced using smaller spindle whorls, takes longer.) So the fabric for a pair of trousers would require at least two hundred hours, or about a month's labor.[19] And that doesn't include the substantial time needed to wash, dry, and comb or card the wool beforehand, not to mention the weaving, dyeing, or sewing.

With this perspective, we can start to understand why even a garment as simple as a plain Roman toga could be a status symbol. Contrary to the impression left by toga party costumes, the toga was closer to the size of a bedroom than a bedsheet, about 20 square meters (24 square yards). Assuming 20 threads to the centimeter (about 130 to the inch), historian Mary Harlow calculates that a toga required about 40 kilometers (25 miles) of wool yarn—enough to reach from Central Park to Greenwich, Connecticut. Spinning that much yarn would take some nine hundred hours, or more than four months of labor, working eight hours a day, six days a week.

Ignoring textiles, Harlow cautions, blinds classical scholars to some of the most important economic, political, and organizational challenges that ancient societies faced. Cloth isn't just for clothes, after all. "Increasingly complex societies required more and more textiles," she writes.

The Roman army, for instance, was a mass consumer of textiles. . . . Building a fleet required long term planning as woven sails required large amounts of raw material and time to produce. The raw materials needed to be bred, pastured, shorn or grown, harvested, and processed before they reached the spinners. Textile production for both domestic and wider needs demanded time and planning.[20]

That was certainly true of the Vikings' famous ships. A Viking Age sail 100 meters square took 154 kilometers (60 miles) of yarn. Working eight hours a day with a heavy spindle whorl to produce relatively coarse yarn, a spinner would toil 385 days to make enough for the sail. Plucking the sheep and preparing the wool for spinning required another 600 days. From start to finish, Viking sails took longer to make than the ships they powered.

Although sail sizes varied with the ship, the total amounts of cloth—and hence thread—were staggering. In the early eleventh century, King Canute's North Sea empire maintained a fleet with about a million square meters of sailcloth. For the spinning alone, that much material would require the equivalent of ten thousand work years.[21]

We blasé moderns may denigrate portraits of women holding distaffs or working spinning wheels as mere symbols of domesticity and subordination. For our ancestors, however, they reflected a fundamental fact of life: without this constant labor, there could be no cloth.

* * *

All over the globe, ancient people devised ways to make thread using a spindle and whorl. It was a brilliantly simple technology, portable and easily crafted from local materials. In the hands of an expert, it could spin remarkably strong, fine, even thread. An Incan *qompi* tunic, a luxury reserved for the honored elite, featured eighty threads per centimeter, or more than two hundred per inch, for the vertical warp alone. But, as exceptional as its products could be, the hand spindle was also slow. The thread to make a single *qompi* took some four hundred hours to spin.[22]

We might imagine, therefore, that spinners in many places would have come up with faster ways of getting the job done. In fact, however, it happened only in China, the birthplace of silk. Only there did some clever soul figure out how to speed up the process by adding a belt and wheel.

Therein lies a paradox. Silk is the only biological fiber that comes in long, continuous strands, known as *filaments*, as opposed to staple fibers. (Synthetic fibers such as polyester and nylon are also extruded in filaments.) The filament from a single unopened cocoon will extend for hundreds of yards and does not need to be spun the way shorter, weaker fibers do. Yet it was silk thread whose production inspired the first mechanical advance in spinning.

To turn silkworm cocoons into usable yarn, the first step is to immerse them in warm water, which dissolves the gum holding the strands in place. With great delicacy and care, a worker—almost always a woman—strokes out the filaments from two or more cocoons with a brush, chopstick, or finger. The strands fuse into a single thread, which she feeds onto a large four-sided reel that a helper steadily turns to unwind the cocoons as they bob and spin in the water. The more homogeneous the filaments, the better the resulting thread. When the silk from one cocoon runs out, the worker picks up the end from another and merges it with the continuous thread.

To keep each new rotation of the wet, slightly sticky silk lying flat and separate from the others, the thread must be stretched horizontally onto a

Silk reeling, portrayed on eighteenth-century Chinese wallpaper made for the Govone Castle in Italy's silk-producing Piedmont region. Although the scene is traditional, the figures' features have been Europeanized for the foreign audience. (Author's photo)

square reel large enough to hold many hundreds of yards. Once the *reel-ing* is done and the thread has time to dry, it is wound onto bobbins and, if desired, twisted together into stronger, more lustrous yarn. The twisting process is known as *throwing*.

This is, at least, the ideal scenario, the one that produces the valuable yarn that Renaissance Venetians called "true silk." But not every silk fila-ment remains fine and unbroken. "Waste silk," inferior but still valuable, is equally essential to our story. Some of it comes from the cocoons of moths allowed to hatch so they can lay eggs; some is the fuzzy down on the outside of the cocoons; some is left in the pot after the reeling process. Whatever the source, waste silk is far too useful and abundant—in the sixteenth-century Venetian mainland, it accounted for about a quarter of all silk—to simply throw away. It can be carded and spun like any other staple fiber.[23]

Here we find the answer to the paradox: Silk is *both* a filament and a sta-ple fiber. Chinese silk workers sometimes reeled filaments, sometimes spun waste silk, and in both cases had to wind thread onto bobbins. Out of these different experiences came the technology that historian Dieter Kuhn de-clares "the first and only labor- and time-saving device developed for the production of yarn and thread" before the fifteenth century. It was the *spin-dle wheel*, which mechanized the first two stages of spinning, drawing out and twisting the fiber. (A fifteenth-century European invention, the *flyer*, wound the thread onto a bobbin, making the process continuous.)

The spindle wheel's inventor was probably a silk worker from the prov-ince of Shandong, a silk center about midway between Shanghai and Bei-jing. Unlike drop spinners, who depended on gravity, she would have long been accustomed to reeling thread with horizontal machines. She applied the same principle to a spindle. She turned it on its side and laid the rod on horizontal supports flanking the whorl so it could still revolve. Then she ran a belt, possibly simply a piece of string, around the top of the whorl, out to a much larger wheel, and then back around. Inspired by silk strands winding onto a reel, this invention marked the first use of the drive belt, an essential component of many later machines. With a single turn of the big wheel, the smaller whorl spun multiple times.

All this happened, Kuhn argues, in the fifth or fourth century BCE, a millennium before the spindle wheel first appeared in India, from which it eventually spread to medieval Europe. Kuhn offers several types of evidence for the early date: a sharp drop in the number of spindle whorls excavated from sites dating to the Chou (1046–256 BCE) and Han (206 BCE–220 CE) dynasties, suggesting the adoption of a different spinning technology; Han dynasty reliefs showing spindle wheels in action; and a significant increase in the number of excavated silk fabrics with twisted and doubled threads.[24]

But we still don't know when the spindle wheel began to be used specifically for making yarn. It's a versatile textile technology that can serve other purposes as well. It can twist, or *throw*, silk threads together, as suggested by those excavated silk fabrics. It can wind reeled silk onto a bobbin, a process known as *quilling*; written Chinese sources record this use as early as the first century BCE. Or it can spin staple fibers, including waste silk. Kuhn interprets an ambiguous image on a Han dynasty relief as showing the spindle wheel twisting waste silk into thread.

He also suggests a fourth reason to believe the spindle wheel was adopted for spinning no later than the Han dynasty: increasing demand. By that time, Chinese weavers were using treadle-operated looms that could weave as many as three meters of hemp fabric a day. Without adequate yarn supplies, adopting this faster but more complicated technology wouldn't make much sense. Using drop spindles, it would take between twenty and thirty hand spinners to keep such a loom supplied with thread. With a spindle wheel, however, spinners could produce thread about three times faster, cutting that number of spinners to between seven and ten. Chinese textile workers, who were already using the machine for throwing silk and winding bobbins, might well have made the connection.

Whatever its initial purpose, the spindle wheel was a technological landmark. It introduced the drive belt, which was adopted for many other uses. It also demonstrated that mechanical power could significantly speed up the process of making thread, reducing a major bottleneck in cloth production. Centuries would pass before this insight would translate into the machines that changed the world. That story, too, began with silk.

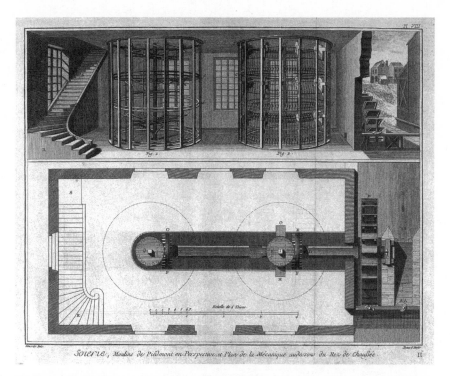

Piedmont silk throwing machines as depicted in the eighteenth-century Encyclopédie *(Wellcome Collection)*

With its twin turrets and balustraded parapet, the Filatoio Rosso could pass as a palazzo. But when it opened in 1678, the imposing edifice was in fact a factory—one of the earliest in Europe. For two and a half centuries, until the 1930s, its skilled workers and water-powered machinery produced silk thread. Today it is the Museo del setificio Piemontese, a monument to the region's silk-producing past. Located in the small town of Caraglio, in northwest Italy midway between Turin and Nice, it houses exact re-creations of the forgotten inventions that gave rise to modern industry.

The museum's star attractions are two enormous circular throwing machines whose whirling operations evoke visions of the Copernican cosmos. Two stories high and made almost entirely of wood, each contains a series of horizontal rings sixteen feet in diameter, supported by columns. The rings revolve around a massive axle that descends to waterwheels hidden in the basement below. Arrayed around the edge of each ring are hundreds of vertical bobbins rotating up to a thousand times a minute. To a seventeenth-century peasant from the Piedmont countryside, it must have seemed like something from another world.

On the first machine, nearly invisible silk strands twist together clockwise, winding upward onto a slightly recessed ring of horizontal spools. The second machine redoubles the yarn, twisting the strands together counterclockwise to make them stronger and more lustrous. Instead of bobbins, its inner circle holds X-shaped reels two feet on a side that wind the silk into skeins. This final product is the warp yarn known as *organzino* in Italian, or in French and English, *organzine*. The doubling is important, because warp yarn has to be tough; warp threads are constantly pulled tight, and the mechanical stresses of a loom's action can easily break them. Weft threads, which cross them horizontally, can be weaker. (To keep the two terms straight, remember that *weft* goes from *left* to right. The old term *woof*, which is rarely used today but often found in literature, is a synonym for *weft*.)

Close-ups of a surviving 1818 silk throwing machine in the Civico Museo Setificio Monti, Abbadia Lariana, Italy (Author's photographs)

Impressive to twenty-first-century eyes, the technology was awe-inspiring in its day. Writing in 1481, the Bolognese humanist Benedetto Morandi took pride in his city's industry, praising the twisting mills that ran "without human assistance except to watch over the silk." In a twelve-hour day, a silk worker throwing by hand might produce a single spindle of thread. A water-powered machine, by contrast, could fill a thousand spindles, with just two or three minders to keep the base lubricated and to repair broken threads. "It was an enormous jump in productivity," says Flavio Crippa, who oversaw the Filatoio Rosso reconstructions. The throwing machine, he declares, "was godmother to a momentous structural change that has gone largely unnoticed."

Trained as a physicist, Crippa spent his career in the modern silk industry, developing and patenting advanced machinery. Over the past two decades, he has focused his considerable ingenuity on recovering and restoring the lost technology of the past. The Filatoio Rosso is one of the many museums throughout Italy that testify to his efforts. Although the building was badly damaged during World War II, Crippa was able to calculate the position and height of its machines by examining the surviving footprints, with a "maximum error of two or three centimeters," he says. With the advantage of modern tools, he laughs, the re-creations took two years to build—just like the originals.

Despite their Bolognese origins, the hydraulic throwing mills found their true home here in Italy's north—in Piedmont, Lombardy, and the Venetian Republic, where water and raw silk were abundant and organzine was in short supply. During the late seventeenth century, wealthy Italian silk traders and French silk manufacturers invested enormous sums to build some 125 mills at the foot of the Alps. These large factories fed the ravenous looms of Lyon, Europe's silk capital.

Along with their state-of-the-art machinery, the mills *alla bolognese* adopted new organizational structures, bringing all stages of production, from harvested cocoon to final skein, under one roof. "The Caraglio factory became the most complete silk yarn factory ever built," says Crippa. "It was called the Filatoio," or throwing mill, "but in reality it was a Setificio," or silk factory, "because it wasn't limited to twisting threads of silk. It produced the thread from cocoons to throwing."[25] Factories throughout the region adopted the same model.

At a single location, a *setificio* might employ hundreds of workers: expert silk reelers, known as *maestre* (the feminine plural of the more familiar *maestro*) in acknowledgment of their expertise; children to wind the reeled silk onto bobbins; workers to tend the throwing machines; and carpenters and blacksmiths to repair them. The Filatoio Rosso even included an on-site convent, whose nuns fed and housed female workers who came in from distant areas.

Vertical integration replaced the cottage industry of old. No longer did reelers operate in independent workshops. No longer did peasant women take home reeled thread to wind onto bobbins. Only with tight supervision and standardization could factories consistently produce thread tough enough to withstand the rigors of the hydraulic throwing mills without breaking.

Piedmontese factories established consistent sizes for reels, installed uniform metal bobbins, and calculated the optimal sizes and speeds for their machines. They developed a mechanism, called *va e viene* (go and come), to evenly distribute thread onto reels, improving its quality. They began to gauge fineness by the weight of a standard length of yarn—a concept still in use—and employed machines that could rapidly measure out a test sample. With their technology, standardization, and closely supervised labor force, the silk throwing mills constituted "a factory system two centuries before the cotton mills of the Industrial Revolution in Great Britain," writes an economic historian.[26]

Piedmontese mills soon set the European standard for organzine, commanding the highest prices and expanding their facilities to meet increasing demand. The family who built the Filatoio Rosso got so rich selling silk yarn that the king of Savoy made its paterfamilias a hereditary count. Walking through the museum's ground level, Crippa points to subterranean excavations visible through glass flooring. They reveal how the reeling operation doubled from ten stations in 1678, each with a charcoal-fueled basin to keep the water warm, to twenty in 1720. Two women, often mother and daughter, worked at each, the less experienced one winding the reel as the veteran coaxed the fine filaments from the cocoons.

Compared to some nearby rivals, the three-story Filatoio Rosso was of modest size. A year before it opened, French merchants built a six-story factory in Racconigi, about an hour's drive northeast, employing 150 workers. Four years later, they added a second plant with eleven stories and three hundred workers. By 1708, little Racconigi had nineteen silk mills, employing a workforce of 2,375 people.

Management, measurement, and machines weren't the whole story, however. The *maestre* were as essential to the success of the mills as the high-tech equipment. They could discern the tiniest differences in fiber size, matching the naturally varying filaments as closely as possible to keep the thread homogeneous and strong. Piedmontese *maestre* also developed a unique technique for crossing two filaments from different baths to wring out the water, making the thread more resilient and rounded. Unlike their counterparts elsewhere, they worked only these two filaments at a time, producing the finest thread on the market. To reward quality over quantity, *maestre* were paid by the day, not by the amount of thread they produced.

It was demanding, highly skilled work, requiring concentration, experience, and constant improvement. Before they graduated to become *maestre*, the young women turning the reels spent years watching the process, absorbing the tacit knowledge of how to handle the delicate filaments. "Rules, patterns of gestures and all the manual automatisms that comprised the art of reeling were gradually transmitted from spinners to reelers during a long period of low-paid apprenticeship," writes a textile historian. This rare expertise was hard to duplicate, making the *maestre* sought-after employees who commanded higher wages than male laborers.

In 1776, when Spanish entrepreneurs established a silk factory in the town of Mercia, they recruited a Piedmontese *maestra* named Teresa Perona with an offer that included a job for her husband, a "trailing spouse" in today's lingo. Her job was more demanding than his, and she worked seven days a week to his six. But her pay was 50 percent higher.

In what was still mostly a peasant society, the *maestre* were industrial aristocrats. In the mid-eighteenth century, the Hapsburg government financed a huge complex in the town of Goriziano, near what is today the border between Italy and Slovenia. Like the Filatoio Rosso, it was a largely self-sufficient campus, including residential quarters and a chapel. The good pay and previously unknown "benefits" drew workers from near and far. The *maestre* were so well compensated that locals took offense. When a group of *maestre* walked through town wearing silk kerchiefs, jealous residents pelted them with stones, forcing the authorities to intervene.

The water-powered silk works of Northern Italy, economic historian Claudio Zanier argues, fostered "a very large female workforce perfectly apt to be adapted to future industrial needs," something he also identifies within the Japanese silk industry. In the nineteenth century, the regions where silk-twisting factories had been concentrated became Italy's industrial heartland—a status they maintain to this day. "The produce of such plants, beside legions of specialized artisans, was a huge disciplined workforce, used to working seven days a week on continuous shifts and in charge of highly demanding quality products," Zanier observes. "They all constituted the necessary prerequisites for an efficient modern factory system."[27]

Yet for all their technological and organizational achievements, Italy's water-powered silk mills rarely rate a mention in accounts of how the West grew rich. "By 1750, across the Alpine front in Northern Italy there were about four hundred water-powered mills. There were more water-powered mills than in Lancaster in 1800," says historian John Styles. "So why wasn't that the Industrial Revolution? Because silk was a luxury."[28]

You wouldn't power a ship with silk sails, package goods in silk sacks, wrap wounds in silk bandages, furnish cottages with silk curtains, or dress laborers in silk garments. (Even in China, which used silk for soldiers' uniforms, ordinary people wore hemp.) As long as mechanical innovations affected only the fabric of a small elite, as prestigious and profitable as it might be, their economic significance was limited. Spinning the staple fibers of ordinary life—wool, linen, and increasingly popular cotton—remained an all-consuming task. But in mechanizing the production of thread, bringing it out of cottages and into factories, the throwing mills foreshadowed the Industrial Revolution.

〽〽〽〽

In 1768, the English town of Warrington, halfway along the River Mersey between Liverpool and Manchester, had largely recovered from the economic downturn that accompanied the end of the Seven Years' War. Although demand for its sailcloth wasn't quite as brisk as during that global conflict, it had rebounded enough to keep three hundred weavers employed. Another hundred fifty wove rough fabric for sacking.

The weavers constituted a tiny fraction of the total textile workforce, however. To supply a single weaver with yarn required twenty spinners—a workforce of nine thousand scattered through the Cheshire countryside. "The spinners never stand still for want of work; they always have it if they please; but weavers sometimes are idle for want of yarn," wrote the agronomist and travel author Arthur Young, who visited the town on a six-month tour of northern England.

Later in the trip, Young journeyed uncomfortably along a turnpike "cut in continual holes" and finally arrived in Manchester. There, he found a prosperous textile industry producing goods for both domestic consumption and export to North America and the West Indies. Jobs were plentiful. "All in general may constantly have work that will," he recorded. In addition to the many workers making textiles, hats, and small wares such as trimmings and tapes, he noted, "the number of spinners employed in and out of Manchester is immense." Thirty thousand spinners worked in the city proper, with another fifty thousand on the outskirts.

In Young's day, spinning was by far Britain's largest industrial occupation. "Taking wool, linen, and hemp spinning together," estimates an economic historian, "the potential employment by 1770 could have been in the order of 1,500,000 married women," in an English workforce of about four million. (This calculation assumes married women spin less than single ones.)

Spinners' wages were modest at best. The Warrington women and girls spinning flax for sailcloth earned a mere shilling a week if they worked full-time, compared to nine shillings for a male weaver or five for his female counterpart. In the Manchester area, adult cotton spinners could earn two to five shillings a week, while girls got between one shilling and a shilling and a half. By comparison, weavers made between three and ten shillings, depending on the type of fabric.[29]

Spinners, it seems at first glance, got a raw deal. "Despite her essential role in England's economic fate, the spinster received penurious wages for her work," writes historian Deborah Valenze. She blames the low pay on sexism. "Stigmatized by its association with women's work, spinning never earned wages commensurate with the demand for thread."[30]

The simple morality tale of oppressed female workers misses the inescapable mathematics of fabric production. Thread may be essential but, unless the final cloth is extremely expensive, the value of an hour of spinning is necessarily low. The *maestre* were well paid, with higher wages than many men earned, because the fabric their work supplied was costly silk. Valenze has the causality reversed. Spinning paid little not because women did it but because it took so long to produce useful amounts of yarn. The product of an hour's labor simply wasn't worth that much. Women took this low-paid work because they had fewer options than men. The oppression lay not in the wages paid for spinning but in the absence of alternative employment for women.

To those in the fabric trade, in fact, spinning wasn't cheap—even at "penurious wages." It easily cost more than the other steps in cloth production. In 1771, a parliamentary report recorded the costs of making a standard piece of worsted woolen cloth selling for thirty-five shillings. The biggest expense was the raw wool itself, at twelve shillings. Spinners' wages closely followed: eleven shillings, eleven and a half pence. (There were twelve pence in a shilling.) Weaving was half as much—just six shillings. The manufacturer earned a profit of two shillings, five pence.

Nor was this ratio an anomaly. For heavy woolen broadcloth, the cost of spinning often ran double the cost of weaving. In 1769, when times were good, producing enough yarn for twenty-five yards of cloth cost seventeen shillings, eleven pence, more than double the eight shillings, nine pence for weaving. Five years later, when broadcloth prices were down, the ratio was even more lopsided: spinners got fifteen shillings, nine pence, while weavers earned seven shillings.[31]

Lousy wages and high spinning costs reflected the fundamental economics of preindustrial fabric production. Cloth required enormous amounts of yarn, and spinning it consumed huge amounts of time. Fine, tightly twisted, consistent thread took even longer. Feeding the looms for anything other than the most luxurious materials was bound to pay poorly. Otherwise, no one could have afforded to buy the fabric.

Spinning was the bottleneck in fabric production, and a bottleneck is a problem waiting to be solved. Beginning in the late seventeenth century, inventors started looking for ways to get more thread with less labor. Like cheap, clean energy today, spinning machines seemed obviously desirable. In 1760, Britain's Society for the Encouragement of Arts, Manufactures and Commerce offered prizes for "a Machine that will spin Six Threads of Wool, Flax, Cotton, or Silk at one time, and that will require but one Person to work it."

Nobody won, but within a few years James Hargreaves introduced the spinning jenny, a horizontal machine that promised to "spin, draw and twist sixteen or more threads at one time by a turn or motion of the hand and the draw of other." It was what economic historian Beverly Lemire calls "the first robust machine that could consistently produce multiple spindles of thread from the effort of a single spinster." Ideally suited for home production, even by children, the jenny sped up spinning, improved consistency, and increased the supply of yarn. More yarn in turn allowed greater production of woven fabric and knitted stockings.[32]

But quantity wasn't the only problem facing British textile manufacturers. With its short fibers, cotton is difficult to spin. Whether using jennies or old-fashioned spinning wheels, English spinners couldn't create cotton yarn tightly twisted enough to serve as warp thread that would hold up under constant tension without breaking. Drop spinning the short staple took so long it was prohibitively expensive. As a result, English "cottons" were actually a rougher cloth called *fustian*, with cotton only in the loosely twisted weft threads and linen in the warps.

What customers really wanted were fashionable all-cotton prints from India, where the spinners were the world's best with cotton. At the behest of Britain's powerful wool industry, however, Parliament had banned Indian imports and, until 1774, even forbade English manufacturers from offering their own all-cotton prints, known as *calicoes*. The British East India

Company was selling ever-increasing amounts of Indian cloth to the North American colonies, where it was far more popular than fustians. English textile manufacturers wanted a piece of the American action. To get it, they needed not just more but better cotton yarn. Spinning, observes Styles, "is not just a bottleneck but the sine qua non of quality."

In a roundabout way, the solution came from the Italian throwing mills. The story begins with a case of industrial espionage, a common occurrence in the history of textiles. In the early 1700s, an English mill owner named Thomas Lombe sent his mechanically talented younger brother John to Italy, in hopes of learning the secrets of Piedmontese silk throwing. Bribing a priest to help him, John secured a job as a mechanic in a Livorno silk factory. By day, he committed the machinery to memory; in the evenings, he put the plans on paper to be smuggled home in bales of raw silk. He returned to England in 1716, bringing a few Italians—and their expertise—with him. Using the pirated plans, the brothers built a five-story silk throwing mill in the town of Derby. It opened in 1722. John died the same year after a long illness reputedly caused by poison from an Italian assassin.

Happy to reward a British subject for importing state-of-the-art technology, however illicitly obtained, the government granted Thomas a patent for the machinery designs. When it was due to expire in 1732, he petitioned for an extension. Parliament instead awarded him the spectacular sum of

Spinning on an Indian charkha, c. 1860 by Kehar Singh (The Cleveland Museum of Art)

£14,000—at this time, an annual income of £100 qualified a family as middle class, £500 as rich—on the condition that he release the plans and a "perfect model" of the throwing machine so that others could copy it.[33]

Shortly thereafter, an inventor named Lewis Paul, the well-connected son of a refugee French physician, began applying the machine's principles to making cotton yarn. Substituting mechanical prowess for human skill, his device employed a series of rollers, each spinning faster than the previous one, to draw out and twist combed and carded fiber into thread. "Circular, with a central drive shaft, it had striking similarities in design to Lombe's Italian silk throwing mill," writes Styles. Paul licensed the technology to investors he met through his friend Samuel Johnson, the prominent writer.

Adopted by factories in northern England, including one in Northampton that installed five machines, each with fifty spindles, Paul's device had technical problems and was only modestly successful. (The mills also suffered from management problems.) But roller spinning inspired other tinkerers. "Several gentlemen have almost broke themselves by it," admitted one of the determined number, a Lancashire barber, wigmaker, and pub owner named Richard Arkwright. Despite his unlikely background, Arkwright was a genius at building on the inventions of others, and he managed to find a solution. Rather than use a circular frame, he lined up several pairs of rollers and, substituting for a spinner's fingers, weighted the top ones to hold the fiber tight, so that the twist couldn't run backward up the draft. The result was consistent yarn with a twist tight enough for warping.

In 1768, Arkwright moved to the stocking-knitting center of Nottingham, enlisted a couple of business partners, and filed for a patent on what came to be known as the water frame. His first spinning mill opened in 1772, with its yarn going into knitted stockings and all-cotton calicoes for the American market. The partners then successfully lobbied Parliament to repeal its calico ban, making the fashionable material, now made with British thread, legally available throughout the country. The water frame, Styles writes, was the "ultimate macro-invention"—a technology that begets others, with consequences that reach far beyond a single function.[34]

Within a few years, water-powered spinning mills had spread throughout northern England, turning out previously inconceivable amounts of low-cost cotton yarn. Over time, Arkwright refined mechanical spinning with water-powered innovations that improved thread quality and integrated carding and roving (twisting fibers to prepare them for spinning) into a single process. He eventually added steam power to his mills. The men who tended his machines, Lemire writes, became "the first generation of elite industrial workers. They were well paid and worked with technologies that brought them considerable prestige."

Nor were they the only winners, at least in the short run. In 1788, Samuel Crompton developed the spinning mule, so called because it combined aspects of Arkwright's design with the bobbins of the spinning jenny. (A mule is a cross between a horse and a donkey, and a female donkey is called a jenny.) The mule for the first time allowed British manufacturers to produce thread as consistently fine and strong as hand-spun Indian cotton. Thread output soared so much that weavers became the new bottleneck.

"Those within the handloom weaving trade enjoyed a golden heyday," writes Lemire, "with as much work as they could want and at high rates of

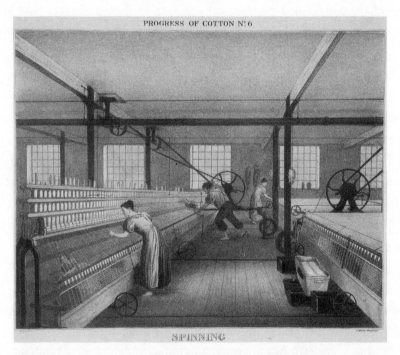

Nineteenth-century spinning mill (Credit: Yale University Art Gallery)

pay." The golden heyday was not to last. Power looms arrived at the turn of the century and with them the fabled Luddite movement, as yesterday's winners became the new economic losers. In one of history's ironies, the hand weavers who broke the looms that threatened their jobs—and became synonymous with resistance to new technology—owed their imperiled livelihoods to earlier, and far more disruptive, advances.

Indeed, Arkwright's earlier generation of "patent machines" sparked its own anti-technology backlash. Protesters smashed machinery and demanded government relief. Pending parliamentary action, the town of Wigan halted the "use of all Machines and Engines worked by Water or Horse, for carding, roving, or spinning of Cotton." A petition to Parliament explained that the "Evil in Question is the Introduction of Patent Machines and Engines, of various Descriptions, which have superseded Manual Labour to such a fatal and alarming Degree, that . . . many Thousands . . . with their Families, [are] pining for want of Employment."

Parliament commissioned a report but decided against action. "A very valuable Manufacture of Callicoes has been established in the said Country by the Use of the Patent Machines," it concluded. For all the disruption it caused, the new technology was creating new kinds of jobs and benefiting the nation as a whole.

A pamphlet with the unwieldy but clear-cut title *Thoughts on the Use of Machines in the Cotton Manufacture. Addressed to the Working People in That Manufacture and the Poor in General* laid out a case that could be made about music streaming, self-driving cars, drone deliveries, or any other fear that *the robots are taking our jobs.*

> Those who were thrown out of their old employments, will find, or learn new ones. Those, who now get less by their labour, will be aiming at the more profitable branches. Those who, by striking early into new inventions, get a disproportionate gain, will soon find so many rivals, that they must sink their terms, and reduce their profits. . . . In fact, the cotton manufacture is almost a new trade. The fabrick, the quality of the goods we make, is amazingly changed. How many new kinds of cloth are made, in very great

quantities, which could not possibly, have been made, at least in any quantity, or so cheap to sell, without our machines?[35]

Though perhaps too sanguine about the immediate fates of individuals, the tract writer was correct about the big picture. By making thread plentiful, the "patent machines" changed the world. From clothing to sails, bed linens to flour sacks, essential items were suddenly much cheaper, more varied, and more easily obtained. Women were liberated from their spindles and distaffs. It was the beginning of what economic historian Deirdre McCloskey calls "the Great Enrichment," the centuries-long economic takeoff that lifted global living standards. Just as string allowed early humans to conquer the world, abundant yarn created ripple effects within nearly every aspect of life.[36]

▨▨▨▨

With a population of about ten thousand, Jefferson, Georgia, is straight up Interstate 85 from Atlanta, an hour northeast of where suburban sprawl turns into woods and pastureland. Until a few decades ago, such small southern towns furnished much of the world's yarn and cloth. On the ever-turning wheel of textile fortune, they succeeded the mill towns of New England and northern England. Now most of the region's textile factories are closed, replaced by newer plants with cheaper labor in China and Southeast Asia. Only the toughest competitors remain.

Fresh from a high-tech textile trade show in Atlanta, I've come to see one of that surviving elite, Buhler Quality Yarns Corp., which styles itself "the leading supplier of fine-count yarns in the United States." Spun from long-staple Supima cotton (a trademark for long-staple, American-grown *Gossypium barbadense* of the variety known generically as pima), its cotton yarn has about 30 percent more fibers than standard upland cotton (the *Gossypium hirsutum* that dominates world markets). Longer, more abundant fibers make the final fabric softer, shinier, and less likely to tear or form pills. But those advantages come at a premium price. "It's either the very best or we can't do it," says David Sasso, Buhler's vice president of sales. For customers looking for the lowest price, someone else will always win.

For my visit, I've dressed in an eight-dollar T-shirt that ordinarily wouldn't make it into a business meeting, no matter how informal the setting. Today, it's a tribute to my hosts. The shirt's super-soft blend of long-staple pima cotton and modal, a luxurious cellulose-based fiber, likely came from this mill. Buhler supplies the big-box store where I bought it. "This is the best value on the market," Sasso boasted the previous day, showing me a T-shirt with the same price and brand. "This is the price of two Whoppers. This is how an efficient supply chain works: eight dollars, with the most expensive fibers in the market."

In the worldwide spinning business, Buhler is a small operation. Its single-story building, a largely windowless specimen of mid-twentieth-century industrial architecture whose light-brick exterior belies the local red clay, houses a plant space with thirty-two thousand spindles. The mill employs 120 factory workers on four shifts.

Although there must be thirty people around somewhere, the factory floor looks almost empty. In the bale room a forklift driver lines up five-hundred-pound bundles of California cotton. A machine arm the width of two bales moves steadily along a row of thirty, skimming off layer after layer of fiber and sucking it into an overhead channel. From there, the fiber goes to a cleaning cycle and then on to carding, combing, and repeated stages of twisting.

Every step is almost completely automated. One of the few humans visible is a woman in an orange T-shirt and denim shorts—cotton demands a warm, humid environment—collecting bobbins as facing rows of six hundred spindles each fill them with thread. A supervisor with a walkie-talkie on his belt and orange earplugs draped around his neck walks the floor. The place is noisy, though not deafening. My T-shirt picks up some debris, but most of the fibers get filtered out of the air by a vacuum system.

Spinning has been around so long that it's easy to imagine the technology is fully developed. That's not the case. "If you look at some of the modern mills today, the number of people in the plant hasn't changed" over the past decade or so, "but the output is a multiplier of maybe two to three times," says Sasso. He shows off a new system called air-jet spinning. Rather than twisting cotton into yarn, it shoots air over the surface of the cotton, wrapping the outer fibers at an angle around the outside. The new machines are quieter than their predecessors and much faster.

"We have 120 people in manufacturing, and we were producing seven million pounds" a year, says Sasso. "We're going to have 120 people in our factory and with the installation of this we're going to be producing close to nine million pounds." That's enough yarn, he calculates, to knit about 18 million women's T-shirts—up from a mere 14 million. Or to put it in terms the spinsters of old might understand, each worker produces about 60,000 pounds of yarn a year with the old machines and 75,000 pounds with the new ones—enough to keep Anna Codde spinning for three centuries.[37]

Chapter Three

CLOTH

The brain is waking and with it the mind is returning. . . .
Swiftly the head-mass becomes an enchanted loom where
millions of flashing shuttles weave a dissolving pattern,
always a meaningful pattern though never an abiding one.

—Sir Charles Sherrington, neurophysiologist,
Man on His Nature, 1940

GILLIAN VOGELSANG-EASTWOOD GIVES EACH OF her six students two bamboo skewers, two colors of yarn, and a small wooden frame with rows of nails at each end. We now have, she declares, "enough to create a loom that works efficiently and is the beginning of the Industrial Revolution. Get on with it."

The problem is much harder than it sounds. Once you've warped the loom by winding yarn back and forth around the nails, it's easy enough to use one stick to hold up every other thread and weave the first row of weft. But then what? If you leave it in, the first stick locks the warp threads into position. How do you lift the second row, and the third? A half hour later, no one has figured out a better way to raise and lower warp threads than just using our fingers.

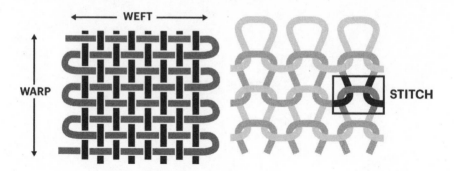

Basic weaving and knitting structures (Olivier Ballou)

Vogelsang-Eastwood, a much-published archaeologist and the founder of the Textile Research Centre in the Dutch university town of Leiden, relishes the big reveal. She ties a loop around every other warp thread, "one, three, five, seven, nine," and runs a stick through the loops, then does the same for the even threads. Raise one stick, pull the weft through, then raise the other and go back. *Voilà.* To make two-dimensional cloth from one-dimensional yarn, you have to think in three dimensions.

In more than a decade of classes, Vogelsang-Eastwood says, only two students have solved the puzzle. One was a weaver who already knew the answer, and the other was an engineer. The ancients who invented the warp-raising loops, known as *heddles*, were "geniuses," she pronounces. We weaving dimwits agree.[1]

Spinning trains the hands, but weaving challenges the mind. Like music, it is profoundly mathematical. Weavers have to understand ratios, detect prime numbers, and calculate areas and lengths. Manipulating warps turns threads into rows and rows into patterns, points into lines and lines into planes. Woven cloth represents some of humanity's earliest algorithms. It is embodied code.

Long before the dawn of mathematical science, weaving brought right angles and parallel lines into everyday life. "Textile patterns don't represent free nature, but they are symmetrically fitted," observes archaeologist Kalliope Sarri. "Weavers could only reproduce motifs . . . if they were able to count, divide, and add up, if they were able to find the center of a circle, the middle of a line, to estimate how many colors to use, how much dye they

Knitting's potential to realize 3-D shapes reflects its mathematical nature. "Every Topological Surface Can Be Knit: A Proof" is the title of a scholarly journal article published in 2009 by topological graph theorist sarah-marie belcastro, who knitted these mathematical objects: a Klein bottle, an orthogonal double-holed torus, and (15,6) torus knots and links. (© sarah-marie belcastro)

needed, and finally to estimate the weight and the economic values of their products." The textile patterns depicted in Neolithic Aegean art, she writes, "reveal the abilities of the weavers in calculating, in conceptualizing and representing geometrical shapes, in creating hierarchies and in estimating sizes, volumes and values."[2]

Knitting, a recent arrival, is equally mathematical, particularly in its ability to create three-dimensional shapes. "Every Topological Surface Can Be Knit: A Proof" is the title of a scholarly journal article published in 2009. "Hidden within almost any knitting project," write two mathematicians who knit, "is not only the arithmetic of counting rows and stitches, but also structural problems which are best understood using abstract mathematics."[3]

"Whether one creates cloth by inventing an elaborate machine, or by constructing the intellectual framework for complex mental computations," observes anthropologist Carrie Brezine, "the existence of cloth is evidence of mathematics at work in the tangible world."[4] At dinner with a group of handweaving enthusiasts, I ask about the relationship between weaving and mathematics. "It's all math," two reply in unison.

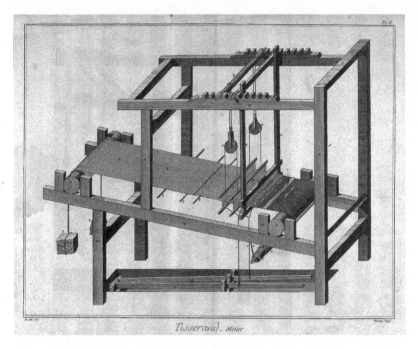

Tisserand, *Metier*.

The earliest cloth was probably netlike, made by looping and knotting string. Later, sewing inspired new techniques such as *nålbinding*, where a blunt needle is used to pass thread through loops wound around the thumb. Although the resulting fabric looks like knitting, the process is quite different. In knitting, a continuous thread is looped to form stitches, with only the loops passing through each other. *Nålbinding*, by contrast, draws the entire yarn length through each loop, using fairly short pieces that are fused together with friction when they run out. Because it doesn't require long, continuous strands, it demands less skill in spinning, and the fabric doesn't unravel when a single loop breaks. Archaeologists have found such cloth in places as far-flung as the Nahal Hemar Cave in Israel and the Tarim Basin in northwest China.[5]

With its interlocking perpendicular threads, weaving represents a conceptual breakthrough, vastly multiplying the number of possible patterns. Although they come in an astounding variety of forms, every loom does two things: it keeps warp threads taut, and it allows the weaver to selectively raise or lower them, creating a *shed* through which the weft can pass.

On the European floor loom, shown on the previous page, in a rendering from the eighteenth-century Encyclopédie, *the warp is wound on beams, and heddle rods are raised with pulleys controlled by treadles. The front of the loom, where the weaver sits, is on the right side of the engraving. Ghanaian kente cloth weavers switch between two sets of heddles, controlled by coconut-shell pedals, to create the fabric's distinctive alternating blocks. Their work requires foreseeing how patterns will interact when the strips are sewn together into a large cloth. (Wellcome Collection, Philippe J. Kradolfer)*

Weaving is the original binary system, at least twenty-four thousand years old. Warp-weft, over-under, up-down, on-off, one-zero.[6]

The possibilities are astronomical. Threads can be loosely woven, tightly packed, or some combination. Warp and weft may be equally prominent, or one may largely cover the other. Threads can be different colors, textures, or materials. Depending on which warp threads you lift, you can alter the appearance and structural qualities of the final fabric. Weaving, says an artist weaver, "is something you can never get to the end of, no matter how long you live."[7]

Instead of two bars of heddles holding odd and even threads, you might use three, lifting threads 1, 4, 7, and 10; threads 2, 5, 8, and 11; and threads 3, 6, 9, and 12. The result would be the diagonals of a *twill* rather than the simple over-under of *plain weave* (also known as *tabby*). Changing the order in which you raise and lower the bars offers more variations, including herringbones and diamonds. Adding rows of heddles further multiplies the number of possible permutations. Color introduces still more.

The smooth surface of *satin* is created by solving a Sudoku-like puzzle: How do you hide the intersections of warp and weft while avoiding obvious twill-like diagonals?[8] Separately or in combination, these three basic structures—plain weave, twill, and satin—can engender countless designs.

Before a single weft thread can cross the warp, the weaver must establish the fabric's structure and pattern. Even plain weave demands forethought: Will it alternate single threads or more? Will there be stripes, checks, or plaids created with different colors or textures? Will the cloth be *double weave*, where the warp threads are selected to create two separate layers? Will the warp and weft threads be equally prominent, or will one dominate the other? Such questions determine what materials you use, how you thread the heddles and space the warp beforehand, and how tightly you pack the weft. With twill and satin, the options multiply further.

"In weaving, art is largely a mathematical thing. It's understanding patterns, it's understanding structure," says Tien Chiu, a "recanted mathematician"—she dropped out of graduate school—and former Silicon Valley project manager now pursuing an artistic career as a weaver. Mathematically, she observes, the threading for a satin poses the same question as what's known as the eight queens problem in chess: Given eight queens, how do you place them on the board so that they share no row, column, or diagonal, thereby preventing any queen from capturing another? For satin, the queens are the places that warp and weft cross, holding the cloth together. Visualizing a weave structure, says Chiu, "for me isn't that different from visualizing abstract algebra."[9]

Today's technologists like to recount how at the turn of the nineteenth century Joseph-Marie Jacquard used punch cards to select warp threads and how his invention inspired Charles Babbage's Analytical Engine, the digital precursor to the computer. "We may say most aptly that the Analytical Engine *weaves algebraical patterns* just as the Jacquard-loom weaves flowers and leaves," Ada Lovelace famously declared. (Emphasis in the original.) It's the one bit of textile history every tech-savvy person seems to know.[10]

Weaving patterns on an Andean backstrap loom requires understanding forms of symmetry. (iStockphoto)

But Jacquard was a latecomer. By the time he invented his card-driven loom attachment, human weavers had been imagining, remembering, and recording complex either-or patterns for thousands of years, harnessing the underlying mathematics.

In the Andes, women traditionally wear a colorful mantle known as a *lliklla*, often using it to carry their babies. Learning to weave the *lliklla* is a rite of passage; the weaver, usually a new mother, graduates to a wider loom after years of making narrow belts. For Ed Franquemont, an American who settled in the Peruvian village of Chinchero in 1976 to learn and record Andean weaving, his first *lliklla* presented a particular challenge.

The cloth typically features patterned stripes alternating with solid bands of color. Producing the design begins with winding a precise number of warp threads in the right sequence of colors—a task performed by an experienced weaver or "warping partner." For Franquemont's *lliklla*, master weaver Benita Gutiérrez wound the warps for designs called *k'eswa* and *loraypu*. Unlike the typical *lliklla* weaver, however, Franquemont hadn't spent years mastering traditional patterns. He knew the *loraypu*, which features an S shape within a diamond, but confessed that he had never woven the zig-zagged *k'eswa*.

"Benita stared at me a moment, and then broke into a broad smile," he recalled.

"'You mean you know *loraypu* but not *k'eswa?*' she asked and began to call neighbors and passersby to share the joke. Soon I was surrounded by a dozen or so laughing women who took delight while Benita's finger traced the *k'eswa* that is within a *loraypu*."

The *loraypu*, it turns out, is formed by joining the *k'eswa* with its mirror image. Franquemont had seen only the final shapes, failing to perceive the essential pattern. He had missed the symmetry, the underlying mathematics of the design—and the key to remembering, reproducing, and embellishing it. For Andean weavers, he later wrote, "learning to weave involves not only mastering the techniques and processes of working the loom, but also dominating the principles of symmetrical operations that build complex structures from relatively simple bits of information."[11]

In an influential 1988 article, mathematician Lynn Arthur Steen argued for a definition of his field more expansive than the traditional "science of space and numbers," with its roots in geometry and arithmetic. "Mathematics is the science of patterns," he wrote.

> The mathematician seeks patterns in number, in space, in science, in computers, and in imagination. Mathematical theories explain the relations among patterns; functions and maps, operators and morphisms bind one type of pattern to another to yield lasting mathematical structures. Applications of mathematics use these patterns to "explain" and predict natural phenomena that fit the patterns. Patterns suggest other patterns, often yielding patterns of patterns. In this way mathematics follows its own logic, beginning with patterns from science and completing the portrait by adding all patterns that derive from the initial ones.[12]

The word *mathematics* refers both to the scientific exploration of patterns and to the nature of the patterns themselves. The symmetries of Andean weaving are mathematical structures. The group theory that describes them is mathematical science. In both cases, "patterns suggest patterns," including "patterns of patterns." Like Molière's M. Jourdain, who spoke prose without realizing it, every weaver is doing math. But as with the motions of the planets, it may take a mathematical genius to first identify and describe the abstract patterns.

As a child growing up in West Germany, Ellen Harlizius-Klück was fascinated by textiles and mathematics, the aesthetic and the logical. To combine her two interests, as a college student she decided to study art and math, with an eye to becoming a teacher.

One of her courses focused on Euclid's *Elements*, the classical text of mathematical definitions and proofs famous for its geometry. When each student had to present a section, Harlizius-Klück got the least glamorous assignment: arithmetic. "I said, 'Oh no, please I would like to do the geometry. Who is interested in arithmetic?'" she recalls.

The professor's response stuck with her. The arithmetic, he told her, is the foundation for Euclid's logical system of proofs. It's less recognized, he said, only because historians can't figure out why it developed. (Many of its ideas are believed to predate Euclid, who wrote around 300 BCE.) Geometry has obvious real-world applications. But the arithmetic seems to simply play games with numbers: "If as many odd numbers as we please are added together, and their multitude is even, then the sum is even," or "If an odd number is relatively prime to any number, then it is also relatively prime to double it."[13] For all its beauty and rigor, this part of the *Elements* seems unmotivated. Why were early mathematicians so interested in what makes numbers odd, even, or prime? Why did they care so much about whether numbers share common factors?

"It's like numbers can be friends or can be relatives," says Harlizius-Klück, "and the reason for that has also to do with the generation of these numbers. So prime numbers have no relatives and friends." *What was that about?*

Mathematicians from Plato to the present have considered ancient Greek arithmetic pure science, inspired only by its own internal logic, with no external stimulus. Harlizius-Klück was skeptical. So was her mathematician husband. "In antiquity," she says, "a mathematician doesn't just invent something like that and write books about it and everyone says, 'Oh how nice.'

We were absolutely convinced that there is something in the background." But they had no idea what it could be.

Then, in the late 1990s, nearly two decades after her Euclid class, Harlizius-Klück took up handweaving. It gave her an idea. "I realized that in weaving, when you want to generate a geometric pattern like a square or a rectangle or a circle, you always have to transform that into arithmetic first," she says. "Because you have to think in *thread counts*."[14] Weaving is all about odds and evens, ratios and proportions—just like ancient arithmetic. Unlike painters, weavers don't draw patterns. They build them up thread by thread and row by row, as if they were creating pictures with pixels on a screen. To do so, they have to understand the kind of numerical relations found in the *Elements*.

Grasping prime numbers and multiples was particularly important when working on the warp-weighted looms depicted on ancient Greek pottery. Here, warp threads, held taut by weights of clay or stone, hang down from a border across the top. The border can be as simple as a braid with the warps knotted around it. But the ancients used something more impressive. They first wove a narrow piece as long as the final cloth was to be wide. Instead of fitting the weft tightly on both sides of the piece, however, they stretched it long on one side—as long as the final cloth would be. When the border was complete, they rotated it 90 degrees and fastened it to the top of the loom. The long former wefts became the warps of the main fabric.

Weavers don't normally count their weft threads, but for a warp-weighted loom's border, the exact number matters. If the number of weft-turned-warp threads is prime, then no repeated pattern will fit evenly across the final cloth. If, on the other hand, the border itself has a neatly repeated pattern, alternating, say, every eight wefts, then the main fabric can include any multiple or common factor of that repeat: every two, four, eight, sixteen threads, etc.[15]

Weaving was pervasive in ancient Greek society. More than an essential craft, it was one of the culture's defining practices, celebrated in ritual and art. Twenty-seven passages in Homer refer to it, including the famous story of Penelope warding off her suitors by weaving and unweaving the funeral cloth she is making for Laertes. Greek poets often used weaving as

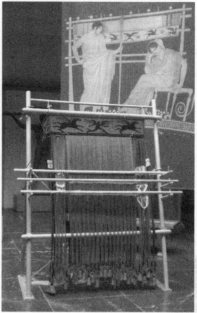

The ancient Greeks' warp-weighted loom as depicted on a lecythos, or oil flask, c. 550–530 BCE, and a full-size reconstruction with fabric made of tablet-woven borders combined with double weave, by Ellen Harlizius-Klück (Metropolitan Museum of Art; © Ellen Harlizius-Klück, 2009)

a metaphor for creating poetry and song. Plato's *Politikos* (*Statesman*), the subject of Harlizius-Klück's PhD dissertation, analogizes the ideal ruler to a weaver, uniting courageous and moderate citizens as the loom joins the strong warp and soft weft. (The dialogue also includes an extended discussion of the stages of wool production.)[16]

At each summer's Panathenaia festival, Athenian women presented the life-size statue of the goddess Athena with a robe, or *peplos*, newly woven in colors of saffron, blue, and purple. Every four years, Athena's monumental statue on the Parthenon also received a new peplos, this one woven by men. The garment was carried to the ceremony mounted as a sail on a full-size ship pulled on wheels and then most likely hung on the wall behind the statue. Woven into each peplos were images of the battle between the Gods and Giants. Small or large, the specially woven cloth, whose presentation is the centerpiece of the Parthenon marbles now in the British Museum, symbolized the unity of the Athenian polis.[17]

So, although speculative, it's reasonable to imagine that early Greek mathematicians knew enough about weaving to be inspired by its logical operations, as they likely also were by land surveying, to take a geometrical example. The more you look at the *Elements* arithmetic through the lens of weaving, the more likely the connection seems to be.

Take the first proposition of Euclid's Book VII: "When two unequal numbers are set out, and the less is continually subtracted in turn from the greater, if the number which is left never measures the one before it until a unit is left, then the number is relatively prime." What this means, in more familiar language, is that if you keep subtracting the smaller number from the bigger one until you wind up with a remainder less than the smaller number, you can confidently say that the larger number can't be divided evenly by the smaller. This reciprocal subtraction has been called "the grand-daddy of all algorithms."[18] It has applications in computer programming, but why would the ancient Greeks have bothered with it? Weaving offers a possible answer.

Suppose you'd like to make a diamond twill pattern that repeats every 19 threads. Your cloth will be about 40 inches wide, with 25 warp threads per inch, giving you 1,000 warp threads. Will the pattern fit evenly? If not, how many more warp threads would you need? Using today's Hindu-Arabic numerals, it's easy enough to do the calculation. But the Greeks had a much clunkier numerical system based on the alphabet. To get a sense of the problem, try using roman numerals to divide M by XIX.

For a weaver, the easiest way to figure out the answer is to bundle 19 threads at a time, alternating sides of the warp, until you end up with a remainder of 12 threads in the middle. You can then add the missing 7 warps to the end of the row, making the total divisible by 19. Despite a convenient decimal system—not to mention electronic calculators—handweavers today still do such hands-on division-by-subtraction because it makes tactile and visual sense. The practice is an example of Brezine's "evidence of mathematics at work in the tangible world." But translating such an everyday, practical method into an abstract generalization required a leap of scientific imagination.

Most woven cloth, today and in the past, is plain weave. It requires planning and care in setting up the loom, especially if multiple colors are involved, but doesn't demand significant concentration. Patterns are more taxing. Whether simple twills or elaborate brocades, they use more complex threading arrangements, often involving many different *shafts* of heddles. As the weaving proceeds, the pattern forces the weaver to think about what comes next and to remember what came before. Let your mind drift and you could lose the thread. The more complicated the pattern, the more essential the weaver's ongoing attention—and the greater the challenge of remembering what to do.

To record patterns, contemporary handweavers use graph paper, sometimes supplemented by computer programs. (Industrial looms are computer-driven.) The diagram, or *draft*, includes four parts: a graph of how the heddles should be threaded; a schematic of the final cloth (dark squares for warp threads on top, white squares for weft); a graph indicating which shafts should be raised together and, if the loom has treadles, tied to the same pedal; and a graph showing the order in which those combinations should be used for each weft row, or *pick*.

Affordable paper is a relatively recent resource, however, and complex woven patterns go back millennia. Through the ages, weavers have developed a variety of mnemonics and storage technologies.

One common approach is to memorize modular patterns, their relations, and the rules for creating them, as Franquemont learned to do in Peru. These algorithms are what Brezine has in mind when she refers to "the intellectual framework for complex mental computations." An Andean weaver can look at a partially woven piece, spot mistakes, and tell what should come next.

Some cultures, from Homeric Greece to contemporary Afghanistan, have encoded thread counts in songs and chants. In *The Odyssey*, Circe and Calypso sing as they weave; hearing her song, Polites knows Circe is at her loom even before he sees her, suggesting the songs were common knowledge. Nineteenth-century rug weavers in central Asia, a European traveler recorded, chanted "in a weird sing-song the number of stitches and the color in which [new patterns] are to be filled."

In Afghanistan, weavers inspired by US propaganda leaflets incorporate pictures of airplanes, the Twin Towers, and American flags into "war rugs." Asked how she translates such images into new patterns, a weaver replied, "I don't see it as a picture. I see it as numbers and I make it a song." In Indo-European cultures, suggests an archaeologist, "counting systems and sing-songs associated with the production of patterns in textiles may have been an extremely early influence on, if not source of, rhythmic or metrically constructed narratives."[19]

The most common storage medium is cloth itself. Surveying the traditional looms still used in eastern Asia, researchers often observed weavers consulting old textiles whose patterns they wanted to duplicate. A weaver in southwest Hunan province attached a reference swatch to the front of her loom even though she said she knew the counts by heart.[20]

Portable and often traded, cloth can carry patterns from place to place. Weavers are "visually trained people with strong numbers skills. They do not need to have physical contact with weavers from other areas to adopt features of different cloth traditions; they only need to have access to textiles from other areas," observes a textile scholar who studies West African weaving, including the cross-fertilization that produced kente cloth.[21]

With enough time, experienced artisans can even decode ancient fabric. Working from archaeological textiles, Peruvian weavers have re-created the finely woven garments of the Juanita Mummy, an Incan girl killed as a religious sacrifice in the late fifteenth century.[22] Textile scholar Nancy Arthur Hoskins has analyzed and reproduced the elaborate geometric patterns in the boy pharaoh Tutankhamun's tunic panels, belt, and collar, demonstrating that ancient weavers almost certainly used weft rather than warp threads to create the designs.[23]

With the earliest known examples dating only to Islamic Egypt, about a thousand years ago, knitting is a far more recent craft.[24] Here, too, modern practitioners decode and reproduce archaeological textiles to better understand their construction. Working from photographs, knitter Anne Des-Moines spent two decades fashioning copies of the complexly patterned stockings from the tomb of Eleonora di Toledo, the wife of Cosimo I de' Medici, who died in 1562. Over the years, DesMoines made four or five versions before settling on what she believes is an exact replica. Unlike modern knitters, she realized, the stockings' creators didn't care whether their background patterns were symmetrical. In addition, she says, "you can tell they were made in a workshop because the second stocking is not the same as the first." Each stocking comprises nine panels, but they join up in different places in the back.[25]

Before they undertook their re-creations, DesMoines and Hoskins had been knitting and weaving for decades, honing the expertise needed to reverse engineer complex patterns. Relying on cloth alone is a risky way to preserve patterns. It assumes that each new practitioner has a teacher to translate the embedded code into manual practice. How to create patterns remains insider knowledge.[26]

That's why what Marx Ziegler did was so radical. A master weaver in the southern German town of Ulm, Ziegler was dismayed to see the city's textile merchants turning to suppliers as far away as Holland to satisfy the demand for their famous linens. Ulm's weavers, they complained, couldn't keep up with seventeenth-century fashions for patterned tablecloths, bed hangings, and window curtains.

"It is sometimes held," Ziegler lamented, "that it would be impossible to make something like it in our own land, as though we were not blessed with as much intelligence as other people." Yet he himself had made all kinds of textiles, from fine linens to heavy rugs, and had mastered patterns requiring as many as thirty-two different shafts. Nor did he believe his neighbors lacked enterprise or talent.

The problem, he concluded, was that ambitious Ulm weavers had little chance to learn how to create patterns, because those who had the know-how hoarded their expertise. "There are those who understand the art but are selfishly concerned that nothing should be revealed," he observed. So, breaking with his profession's traditional secrecy, Ziegler decided to write an instruction manual. In 1677 *Weber Kunst und Bild Buch*, or *The Weavers Art and Tie-Up Book*, became the first book of weaving patterns ever published.

It took professional boldness for Ziegler to put his weaving know-how into print. It also required a written code: a way of translating the directions for woven patterns into easily decipherable diagrams. Just as musical scores

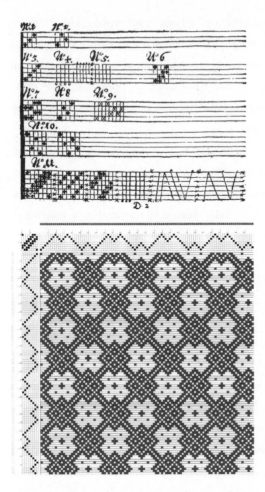

A page from Marx Ziegler's original weaving handbook and a modern weaving draft based on its pattern 11 (Handweaving.net)

record which notes to sound, Ziegler's book used lines and graph paper to show weavers how to thread their looms and which shafts to raise to create specific patterns. His diagrams provided a layer of abstraction between material practice and pure mathematics. Today's weaving drafts descend from its notation.

Ziegler's book expressed the period's growing belief in sharing useful knowledge—the attitude that a century later would culminate in Diderot's massive *Encyclopédie*, with its detailed, illustrated articles on mechanical arts, ranging from wig making to slate mining and including several types of weaving. "I hold that it would be possible to produce many more artists in all branches of knowledge," Ziegler wrote, "if only there were no shortage of publishers."[27] He thus played a role in what economic historian Joel Mokyr has called the Industrial Enlightenment, as scientific theorists and practical craftsmen increasingly joined forces, informing one another's understanding and making both kinds of knowledge more widely accessible.[28]

Weavers had long used private notation to record patterns. Condemning the secrecy of his fellow craftsmen, Ziegler wrote that "tie-up books must be necessary for draw loom and treadle-loom weavers, since without them no one could learn to weave or accomplish anything of the sort." But before his book, those diagrams were trade secrets. Only when its intermediary code became public, and therefore visible to nonweavers, could weaving inspire outsiders to learn from its examples or apply their ingenuity in its formerly closed domain.

Ziegler's book and similar manuals, writes Harlizius-Klück, "made the art of weaving public, and its notation became standardized and common. With the notation close to the machine, they furthermore facilitated mutual understanding of the interaction between pattern drafting and loom parts for nonweavers and through this enabled engineers and inventors to play around with the mechanisms finally leading to an automated loom."[29]

𝍤𝍤𝍤𝍤

Bouakham Phengmixay depresses her loom's treadle, raising the shaft that holds the heddles carrying every other warp thread. With her right hand, she propels a shuttle of red silk through the shed, catching it in her left hand. She grabs the comblike *reed*, whose fine vertical bars hold the warp threads in order, and pulls it forward, pushing the weft into place.

If she kept going like this, Boua, as she's known, would create a fine silk fabric in a plain weave of red weft and black warp. But the textile she's weaving is much more colorful and complicated, with an intricate geometrical pattern in white, red, black, green, and golden yellow. It resembles embroidery but is in fact a *brocade*, in which the weaver inserts *supplementary weft* threads to create a design. Unlike the red and black plain weave, known as the *ground*, the supplementary weft plays no structural role; you could remove it and the fabric would remain intact. Mostly obscured by the brocade, the ground weave holds the material together.[30]

Requiring the weaver to select warp threads one by one, brocades are tremendously demanding to design, plan, remember, and weave. Through the centuries, they have been among the most prestigious and expensive textiles, adorning courts and courtiers from Versailles to the Forbidden City. Yet in Laos, Boua's home, ordinary rural women have long worn silk brocade and used it to decorate their homes. They could afford these ornate fabrics in part because they raised, spun, and dyed the silk themselves. But the real secret was an ingenious technology for creating and storing the pattern code.

At the back of Boua's loom hangs a gauzy web of plain white nylon string. Visually, it's the skeletal ghost of the colorful brocade. Functionally, it's the software that controls the pattern.

Its vertical strings are special long heddles. Unlike the heddles for the ground weave, these aren't on shafts raising the same threads together row after row. Instead, each warp thread can be raised or lowered independently. The heddles are long so that they have room for the horizontal nylon strings that weave in and out across them. Each nylon string runs behind the heddles controlling a single row of supplementary weft, separating those warps from the rest.

After adding her row of red plain weave, Boua reaches out and grabs the ends of the bottom horizontal string. She pulls it forward to separate the heddles in front of it from the rest of the mass, then grasps the selected heddles with her left hand to lift the warp threads they're attached to. With her right hand, she inserts a wide, flat stick known as a *weaving sword*, under the raised warps, between the ground-weave heddles and the long ones. She turns it on its edge to raise the selected warp threads. Her hands now free, she's ready to add the pattern weft.

Her quick fingers pull a silvery strand of white silk over a few warp threads, then under raised warp threads and to the side, laying it on top of the growing fabric. Her hands move to the left, repeating the motion with another supplementary strand, and then another, sometimes introducing a new thread, sometimes extending one up from the previous row. To the untrained eye, it seems like she's tying knots. When she gets to the end of the row, she pulls the reed forward, then turns the sword back to its flat

Lao loom from the back, looking toward the weaver through the pattern strings, and from the side, with the newly created fabric and weaver's bench to the right. In the side view, the weaving sword is inserted between the pattern heddles to the far left and the ground-weave heddle shafts to its immediate right, lifting selected black warps. The reed is at the very front. (Author's photos)

position. Where the warps were raised, we see the slightly recessed ground of black and red; the colors fill in the rest. She depresses her treadle and adds a new row of red, building the ground weave that holds the design in place. She can then select a new string from the web, controlling a different arrangement of heddles, or repeat the previous line of pattern by turning the sword back on its side. The colors are up to her.

Whether copied from old fabric or designed from scratch, the string template for a Lao brocade can take months to create. For each row of supplementary weft, a master weaver like Boua first picks out the desired warp threads, lifting them with a pointed stick. She inserts a looped string behind the long heddles tied to those threads, hanging it on posts on either side of the loom, then repeats the process for the next row, stacking the strings vertically. Like programming a computer, the process requires skill and concentration. But once the template is finished, any weaver familiar with the loom can use it. When a weaver wants to change patterns, she can roll up the old web for storage and install a new one.[31]

Adding decorative weft (or in some cases warp) threads is an ancient practice, going back at least five thousand years to the Neolithic weavers whose work was preserved in Alpine Swiss marshes.[32] In many cases, such as the elaborate Incan weavings done on backstrap looms, weavers picked out the pattern thread by thread as they went. But Lao weavers weren't the only ones to develop a storage mechanism. The Moanan people in China's far northeastern mountains use a barrel-like basket of bamboo pattern rods that rotates as the weaving advances. In southern Guangxi province, Zhuang weavers employ a similar system sometimes called the "pig basket" for its resemblance to the cages farmers use to bring their pigs to market.[33] Like a Lao web, these patterns can be stored and reused.

The most influential mechanism for selecting brocade threads is the *drawloom*, which probably originated in China and spread west, with regional variations developing in India, Persia, and Europe. Unlike most pattern looms, the drawloom was designed for large-scale workshop production of long bolts of luxury fabric, rather than household weaving of pieces a couple of yards long. It could create ground fabrics of satin or twill as well as plain weave. Above all, it was versatile, permitting detailed, high-resolution patterns.

"The Chinese Drawloom is in fact one of the first mechanical devices able to reproduce intricate designs in colour, surpassing woodblock printing of the same period," write researchers Eric Boudot and Chris Buckley, whose work documents traditional looms still used in East Asia.[34] If a Lao loom is *Ms. Pac-Man*, a drawloom is *Grand Theft Auto*.

Think of it as a much larger, two-person version of Boua's loom. Instead of a mesh of strings at the back, the Chinese drawloom has a high tower above the warp with a seat for an assistant, often called a drawboy or drawgirl, and an elevated series of pattern loops. The assistant does the work of pulling and holding the pattern strings, replacing Boua's fist and weaving sword. This arrangement has room for many more pattern strings than a Lao loom and therefore a more detailed and varied design. Using a drawloom, a weaver with enough time and expertise can create almost any pattern.

"The price that is paid for this flexibility," write Boudot and Buckley, "is a complex mechanism and the high level of skill and involvement that is

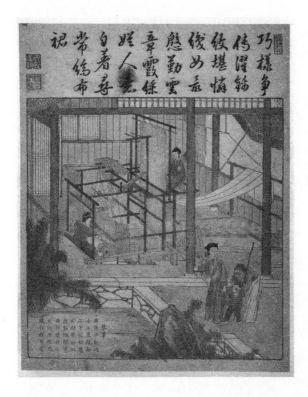

祝 常 自 照 章 憋 後 彼 傳 巧
織 著 人 霞 勤 必 堪 遲 樣
布 尋 老 係 雲 忝 憁 錯 爭

The Chinese drawloom, with the drawgirl controlling the pattern strings from above. (Rare Chinese Books Collection, Library of Congress)

required of both the weaver and her assistant, with frequent pauses to repair broken or tangled cords. Drawloom textiles are correspondingly expensive to make and the system is only used for the most luxurious textiles."[35] These precious brocades adorned altars and priests, palaces and kings. Ordinary people rarely owned drawloom cloth. It was simply too difficult to make.

For the lavish courts of eighteenth-century France, virtuoso designers pushed the drawloom to its aesthetic frontier. In the waning years of Louis XIV's rule, they created gold brocades displaying naturalistic shapes such as exotic pineapples rendered in metallic threads. The patterns celebrated the latest discoveries in natural history and displayed the weaver's ability to simulate curves.

As the decades passed, designs grew lighter and more colorful. Instead of the opulence of gold, they employed more subtle signals of wealth and power. "The designers covered all the eighteenth century with flowers," says Claire Berthommier, head of collections at the Museum of Textiles in Lyon. "Why? Because flowers are very difficult to translate into silk. It is a very

An example of Philippe de Lasalle's ability to create naturalistic designs on the drawloom. Silk and chenille brocaded on silk satin ground, c. 1765 (Los Angeles County Museum of Art)

skilled operation to translate something made with watercolors into something made with thread."[36] By portraying realistic bouquets of multicolored flowers, designers and weavers displayed their skill at manipulating colored threads, giving buyers a new way to demonstrate their wealth and taste. French brocades reversed archaeologist Kalliope Sarri's observation about Neolithic textiles, crafting the illusion of "free nature" within the perpendicular geometries of the loom.

The most accomplished of the century's textile designers was Philippe de Lasalle (1723–1804). Acclaimed as the Raphael of Silk, he was more like a business-oriented Leonardo. Trained as an artist in Lyon and Paris, Lasalle combined painterly skill with a deep understanding of the drawloom's capabilities, an inventor's mind, and a hard-won appreciation for the politics of textiles. In the 1750s, he developed formulas for bright, durable dyes for printed silks, only to have the Lyon Chamber of Commerce discourage his efforts lest the competition from prints hurt the brocade industry.

From then on, Lasalle turned his artistic, entrepreneurial, and inventive energies to brocades. Praising his designs, a contemporary effused:

His fabrics seemed to preserve the natural movement of vegetation with the elegance of a jet of water. The purity of forms, birds, and insects animating his picturesque compositions, and the fresh landscapes showed what our industry could create under the direction of this intelligent artist.

To translate a sketch into thread, a silk designer first created a *mise-en-carte*, a large-scale version on graph paper, with each square representing a single intersection of warp and weft. The challenge was not merely to produce a pleasing picture but also to foresee how the oversized design would translate into fine silk. Too much detail and it would lose clarity. Too little color gradation and it would lack grace. By using similar colors for shading and specifying varied textures of yarn, Lasalle gave his brocades a sense of realism and depth.

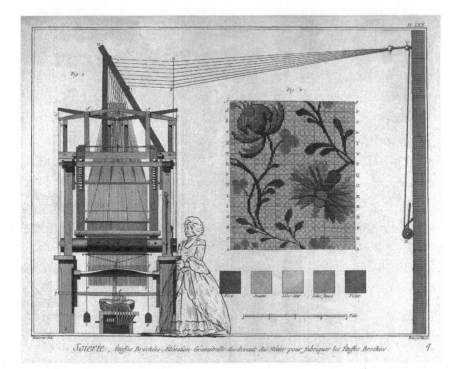

The French drawloom, with its vertical cords, and an example of a mise-en-carte. *The drawgirl (although not dressed as finely) would stand by the side of the loom, as shown. (Wellcome Collection, iStockphoto)*

Once on paper, the pattern had to be encoded for the loom—a time-consuming and finicky process. A silk brocade might have three hundred warp threads to the inch and hundreds, even thousands, of weft picks in a pattern repeat. Setting up a new design could take three months, during which the loom couldn't be used for weaving.

On the French drawloom, the assistant (usually female) didn't sit above the warps. Instead, she stood near the weaver, pulling vertical cords that hung down next to the loom. To set up a pattern, a *liseuse*, or reader, called out the colors thread by thread, while a second worker tied a lash around the corresponding control cords, known collectively in English as the *simple* (or *semple*). Each lash lifted the weighted heddles for a single row of supplementary weft. The drawgirl's job demanded—or developed—both concentration and muscles. Raising hundreds of heddle weights and overcoming the cords' friction required strength, endurance, and on occasion a team of several drawgirls.

The system also had a major commercial drawback. Each stage of cords tied on directly to the next, making it impossible to store patterns for future use. Once a brocade was completed, the knots had to be undone to make room for a new design. If a customer ordered an old pattern, the process had to start again. And the weaver couldn't easily combine two patterns in the same run of cloth. The result was a practical limit on the size and variety of patterns.

Lasalle, who throughout his career devised many loom improvements, decided to attack the problem. After nine years of trial and error, he produced a removable simple that could be prelashed in a given pattern and swapped in and out as needed. A workshop could even prepare simples for the season's new offerings while sales agents were out soliciting orders. In a fashion-conscious era, reducing turnaround time was a major benefit. The removable simple made larger, more varied patterns economically feasible, winning Lasalle further acclaim as both a designer and an inventor.

Lasalle scored a publicity coup with the first woven portraits. He chose royal subjects, including Louis XV and his grandson, the comte de Provence, and signed his work in Latin capitals like the inscription on an ancient Roman building: "LASALLE FECIT." His profile of Catherine the Great hung in the home of Voltaire, gaining Lasalle commissions from the Russian monarch. "The brocade technique in Catherine's medallion is so delicate," writes a historian, "that only by examining the reverse can one be sure it is woven and not embroidered."[37] Lasalle's woven portraits, like his removable simple, foreshadowed the now-famous weaving programs to come.

꧁꧁꧁꧁

The most famous woven portrait ever done was not by Lasalle but by the Lyonnais weaver Michel-Marie Carquillat, who based it on a painting. It depicts a slightly stooped man in a brocaded chair. Surrounded by carpentry tools and loom parts, he holds a drafting compass poised on some perforated cardboard strips. Next to them is a small-scale model of a loom, with a track of punched cards winding like fabric down its back. The sitter is, of course, Joseph-Marie Jacquard.

With fine details, including cracked window panes and a diaphanous curtain, Carquillat's woven scene is far more complex than any of Lasalle's cameo-like portraits. "Only after the Jacquard loom came into use could a work of this extreme level of detail be produced," observes the Metropolitan Museum of Art's website. The picture looks like an engraving. In his memoirs, Charles Babbage recounts that the Duke of Wellington and two members of the Royal Academy of Arts mistook the version in his home for such a print.[38]

Portrait of Joseph-Marie Jacquard, woven by Michel-Marie Carquillat (Metropolitan Museum of Art)

Jacquard's invention made patterns more flexible and easily stored than Lasalle's removable simple. Most important, it entirely eliminated the need for an assistant. Although it built on earlier inventions that used cards or perforated paper for control, it was the first commercially practical design. A self-taught mechanic, Jacquard brought his personal experience as a weaver to bear on the problem. "The merit of Jacquard is not, therefore, that of an inventor, but of an experienced workman," wrote a nineteenth-century observer, "who, by combining together the best parts of the machines of his predecessors in the same line, succeeds for the first time in obtaining an arrangement sufficiently practical to be generally employed."[39]

Jacquard's invention works like this: The cord controlling each weighted heddle catches on the bottom of a double-ended hook. The hook passes through a thin horizontal rod, or *needle*, with an eye in the middle, a spring on one end, and a sharp point on the other. The top hook hangs on a bar above, called the *griffe*. "The whole function of the apparatus," explains the 1905 *Encyclopaedia Britannica*, "is to liberate these hooks in the order and to the extent necessary for the successive sheds."[40]

That's where Jacquard's famous punch cards come in. Opposite the sharp end of the horizontal needles is a perforated rectangular wooden box called the *cylinder*, despite its flat sides—one of Jacquard's refinements over an earlier truly cylindrical design. Each side of the cylinder fits a single card representing one row of weft. The cards for the full pattern are sewn together in a belt.

When the weaver steps on the loom's treadle, the cylinder moves back and rotates to advance to the next card. It then returns to contact the needles. If a needle enters a hole, the hook above it stays put. If there's no hole, the rod pushes back against the spring and this motion slides the hook off its resting place on the griffe. The griffe then rises, lifting the hooks still attached to it and raising their warps. After the weaver inserts the weft, the griffe returns to its place. This automated system allows the weaver to control every warp thread individually without relying on an assistant.

Jacquard didn't invent digital patterns or storage. He invented a way to automate their execution. "The really important idea—the one Jacquard certainly did pioneer—was the notion of applying punched cards in the loom

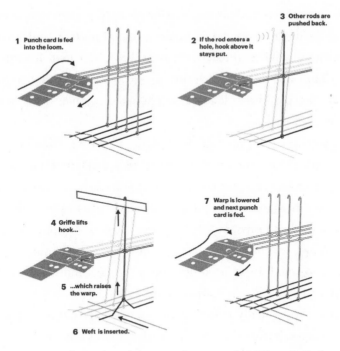

How Jacquard's punch-card-driven mechanism worked (Olivier Ballou)

control system automatically," observes science writer James Essinger, "so that the loom in effect continually feeds itself with the information it needs to carry out the next row of weaving." When Jacquard's device was patented in 1804, Essinger writes, "it was unquestionably the most complex mechanism in the world." To work properly, it required parts engineered to close tolerances.

The complicated mechanism greatly simplified the weaving process, even for fabrics without supplementary wefts. No longer did the loom need multiple treadles to control separate shafts for the ground weave—or for ordinary cloth. A card took care of everything. All the weaver had to do was step on a single treadle to move the cards, pull the string to release the flying shuttle for the ground, and beat the weft into place—a sequence that gave rise to the onomatopoeic Lyonnais word *bistanclac* to describe the loom. The result was an enormous boost in productivity, especially for fancy patterns. A single master weaver could now turn out two feet of brocade a day compared to a mere inch working with an assistant on the old drawloom.

With no technical limit on the number of cards, pattern repeats could be as long as the designer wanted. Take the Jacquard portrait, which measures 55 centimeters by 34 centimeters, or about 2 feet by 1 foot. It used *twenty-four thousand* cards, one for each row of weft, with more than a thousand holes each. A typical brocade repeat might take a tenth as many, each with several hundred holes.

The cards did not program themselves, of course. Making the ones for the portrait took months. But even there, a single *mise-en-carte* reader punching cards on a machine could do what previously required two people, the *liseuse* and her loop-tying partner. Once made, chains of cards could be labeled and stacked easily on shelves, allowing a workshop to produce works to order. When someone like Babbage wanted a copy of Jacquard's portrait, the workshop could weave it on demand. If needed, individual cards for a specific design could also be rearranged or replaced.[41]

Lyon's weavers initially resisted the newfangled apparatus, despite its obvious advantages, fearing the loss of jobs. The protests sometimes turned violent, with the city's *conseil de prud'hommes*, or labor tribunal, breaking up the machines in the public square. Although honored by Napoleon and granted a pension for his invention, Jacquard had to flee the city several times.

But Lyon's weavers eventually embraced the technology. Its versatility gave them a competitive edge over silk manufacturers in England, Italy, and Germany, allowing the city to reclaim the dominance that had eroded since the French Revolution. To accommodate the two-story machines, weavers moved to new, high-ceilinged workshops on the slopes of the steep Croix-Rousse hill between the Rhône and Saône Rivers.

By 1812, a nineteenth-century historian observed, a "true revolution" had taken place in the city's silk industry. Incremental improvements boosted the loom's speed and cut the cost in half. Far from devastating the city's employment, Jacquard's invention ushered in a new golden age of Lyonnais silk. By the end of the century, Croix-Rousse rang with the sound of twenty thousand *bistanclacs*. Jacquard's breakthrough made weavers' work easier, improved the quality of their fabrics, and expanded the market for their products to middle-class shoppers. The card-driven system spread to makers of ribbons, woolens, and carpets.[42]

Showcased at international expositions and adopted around the world, Jacquard's apparatus made weaving code tangible in a way that could inspire nonweavers. Shipbuilders designed similar systems to control the automatic riveting machines used to build the era's new ironclads. Most resonant in our digital age, the binary structure captured the imagination of Babbage and his successors. "Many of the subroutine methods and editing systems that are standard in modern computers were conceived in the 19th century to produce cards for textile patterns," wrote computer scientist Frederick G. Heath in 1972:

> Having a pattern and desiring a binary sequence for weaving it is the same as having a Fortran program and wanting the equivalent binary code that puts the program in terms suitable for a computer.
>
> Indeed, the connection between weaving a textile fabric and designing a computing system is a close one. Anyone who looks at the wiring of a computer or at an enlargement of a large integrated circuit will notice a strong resemblance to normal fabric patterns.[43]

In the first few decades of computing, the connections between the ancient code of cloth and the futuristic promise of information technology took tangible, visible form.

▨▨▨▨

When Robin Kang moved her studio from Queens to Brooklyn in April 2018, it took four months to rethread the 3,520 heddles on her Jacquard loom. Designed for artists, Kang's loom marries computer-controlled warp to hand-inserted weft. It bridges the divide between the old-style Venetian looms that still weave a few centimeters of velvet a day for clients like the Kremlin and the computer-driven industrial looms that produce that much in seconds.

An ebullient blonde from a small town in West Texas, Kang started as a digital printmaker in the 1990s, when Photoshop was a new tool. She learned to weave in graduate school. Like digital image making, weaving spoke to her fascination with algorithms. "It's very computational," she says.

"The weave draft was exactly the concept that I was interested in, in thinking about computer algorithms—you set these parameters and this is what happens." Weaving, she discovered, was the ideal medium to combine her digital fascination with her love of the tactile and the visible artist's hand.

Juxtaposing black warps with weft in metallic threads and vivid hues, her work pays homage to the early history of computing. Take "Lazo Luminoso," a piece dominated by a grid shading from blue to green. A ring circles each intersection, with gold thread wending through it and around the edges, creating heart-like patterns when it crisscrosses the diagonals. The piece looks like a weaving of weaving—and in a sense it is.

The image is inspired by *magnetic core memory*, the dominant computer storage medium for two decades, until the emergence of silicon memory chips in the early 1970s. Each core memory plane comprised woven copper wires with a tiny ferrite bead at each cross, representing a single bit. The washer-shaped beads, no bigger than a pencil point, were called *cores*. Passing a strong enough electric current through a core would create a magnetic field going either clockwise or counterclockwise. Reversing the current, again at a level above a critical threshold, flipped the field. These two states could thus represent 0 and 1. By sending half the current through the "weft," or x coordinate, and half through the "warp," or y, the system could

Magnetic core memory magnified 60 times (Author's photo)

identify and change one specific bead. Wires along the diagonals read the signals.

Kang shows me some examples, including a waist-high array of ninety-six four-inch squares, each with sixty vertical "warp" wires and sixty-four horizontal "wefts." In the early days, women using microscopes hand wove the entire grid. By the mid-1960s, when Digital Equipment Corporation produced the first minicomputers, machines could take care of the x and y wires, but weavers still had to thread the diagonals. "They're literal weavings," Kang says, "and looms too in a way."[44]

For the Apollo space program, a different type of memory held the software.[45] Programmers first wrote the code using punch cards. Once it was finished and debugged, however, it had to be converted into something more resilient and lightweight: *rope memory*. Here, if a wire went through a magnetic bead, it represented a 1; if it went around the bead, it represented a 0. "The software for Apollo was an actual thing. You could hold it in your hand and it weighed a few pounds," writes historian of technology David Mindell.

To produce the rope memory, NASA turned to Raytheon Company, a defense contractor in the old textile and watch manufacturing town of Waltham, Massachusetts. "We have to build, essentially, a weaving machine," a Raytheon manager explained at a news conference. Familiar with both high-precision machinery and textiles, the local workforce was well suited to the task. "You would have to send the program to a factory, and women in the factory would literally weave the software into this core rope memory," says Mindell. Weaving thousands of hairlike wires into a single program took months, but the result, he writes, "was indestructible, literally hard-wired into the ropes."[46]

Although Raytheon evoked looms, the engineers who created magnetic core memory didn't think about weaving. Groundbreaking IBM researcher Frederick H. Dill says he "never heard of 'stringing cores' as a weaving problem" but rather "looked at the wires through the cores in terms of what electrical signals they provided."[47] To embody code, computer pioneers unknowingly imitated cloth. The form of the memory devices arose from the fundamental mathematics of weaving.

〽〽〽〽

After more than ten thousand years of dominance, weaving no longer rules the textile world. Knitting has staged a coup. As I write, everything I'm wearing except my jeans—my underwear, shirt, sweater, socks, even my sneakers—is made from knitted fabric. And I'm a holdout against the leggings and yoga pants that have eroded the denim market. Knits now outsell woven textiles nearly two-to-one, and the ratio for garments is even more lopsided.[48]

One reason is comfort. With the popularity of athleisure styles, the flexibility of knits has triumphed over the crispness of wovens. Rather than sticking to a grid, the thread in a knit winds up and down through the fabric, allowing the cloth to stretch. In earlier periods, the stretch could sometimes result in a misshapen sweater, but today's spandex blends snap back. Knits' forgiving structure makes them easier to fit.

Fashion isn't the only reason knitting has conquered the textile market. With no heddles to thread, industrial knitting machines are much faster to set up than looms. You can swap in new colors or textures in a matter of minutes. And, unlike two-dimensional wovens, knitting lends itself to designing in three dimensions. The first commonly knitted garments were stockings and caps, created with a continuous spiral of stitches passing among four or more needles in what is called "knitting in the round." Late Medieval artworks show the Virgin Mary using the technique to make a seamless garment for the Christ Child.

The Madonna knitting in the round, from the Buxtehude Altar by Bertram von Minden, c. 1395–1400 (Bridgeman Images)

Perhaps inspired by the long, fine wool of the sheep grazing in nearby Sherwood Forest, in 1589, a twenty-five-year-old English curate named William Lee invented a machine for producing stockings. Known as a *stocking frame*, it could knit a row in a single pass, using special "bearded needles" to keep the previous row's loops safe while the new ones interlocked with them.

A century of refinements followed, eventually turning *framework knitting* into a significant form of textile production, existing alongside the hand knitting still valued for its quality and flexibility. By the mid-eighteenth century, Britain had around fourteen thousand stocking frames, each requiring more than two thousand components, including fine needles that challenged the blacksmith's art. A stocking frame was more like a compact floor loom than today's industrial machines. But the basic idea was similar.

The original stocking frames produced plain fabric, using only the *knit stitch*. In 1758, Jedediah Strutt of Derby invented a way to make ribbed stockings, incorporating the contrasting *purl stitch*. These flattering styles boosted the appeal of machine-made hose. Once patented, the "Derby rib"

Framework knitting workshop, 1750 (Wellcome Collection)

The stocking frame required more than two thousand individual parts. (Wellcome Collection)

machine proved highly profitable and played its own role in the Industrial Revolution. When Richard Arkwright moved to Nottingham to establish his first spinning mill, Strutt and his partner financed the start-up. Hosiery was one of the earliest uses for the new cotton thread.[49]

In the twenty-first century, the Derby rib has taken on a new incarnation. Vidya Narayanan holds it up as we tour the Textile Lab of Carnegie Mellon University's renowned Robotics Institute, where she's a grad student. It's a stuffed bunny with a familiar form and an unusual exterior. Known as Stanford Bunny, a standard model for testing 3-D computer renderings, this version is covered ears to tail in what looks like a light-blue ribbed sweater.

Trained in computer engineering and computer science, Narayanan is not a knitter. "Oh God, no," she replies when I ask whether she had any knitting background before arriving in Pittsburgh. She was interested in graphics and fabrication, and Jim McCann, the former game designer who now heads the Textile Lab, convinced her that flatbed knitting machines could act as 3-D printers for soft objects. "Machine knitting is at this very interesting spot," says McCann. "It feels like it's really close to where 3-D printing was maybe at the turn of the century." Industrial machines are well established and models within reach of individuals are on their way.

Most industrial knits are made on fast circular machines that knit continuous spirals, turning out hundreds of yards an hour; the fabric is then cut and sewn to create the final product. Flatbed machines are slow by comparison but far more flexible, because each of their hundreds of needles can operate independently. A relatively modest machine with two beds of seven hundred needles each can theoretically produce more than a trillion different fabrics—even without changing the yarn to vary color or texture. Unlike circular machines, flatbed machines can also create detailed, three-dimensional garments.

On a flatbed machine, says Michael Seiz, a veteran knitting engineer who has worked on both the machine and apparel sides of the market, "I'm probably lucky if I can knit five yards an hour. But I can knit totally shaped pieces. I also can knit an entire sweater in an hour." No cutting, no sewing, no wasted fabric or yarn.

Japanese manufacturer Shima Seiki introduced seamless garment knitting in the mid-1990s, but the apparel industry has embraced the technology

Stanford Bunnies created with 3-D knitting and software developed at Carnegie Mellon's Textile Lab. The one on the left uses the basic stitch, whereas the more advanced one on the right incorporates patterning. (Vidya Narayanan, James McCann, Lea Albaugh)

only in the past decade, as digital machine controls have improved. Using a flatbed machine, a sneaker manufacturer can now make an entire shoe—everything except the rubber sole—in a single piece, varying the knit structure to support the arch, shape the heel, or hold the shoelaces. Assembling the final shoe requires only a little bending and some glue. With this simplified production process, the manufacturer can replace shoe inventories with less risky yarn supplies, knitting whatever models prove to be in demand.

But 3-D knitting still faces a technological hurdle: expensive proprietary software requiring specialized expertise. Like weavers, knitters chart their patterns using square boxes to represent stitches. Those written codes make it possible to easily share designs across time and space. But to interpret the code, you need to know how to knit and, crucially, how to translate the two-dimensional representation into three-dimensional results. That's as true of computer knitting as it is of hand work. "Knit programmers are still doing the same thing they were eighty years ago," write Narayanan and her coauthors. "Though the media has changed—from cam cylinders to card chains to paper tape to floppy disks to flash drives and ftp servers—programmers still explicitly tell the machine what to do with each needle on each carriage pass."

The machines keep getting more refined, with ever-improving needles and controls, but the software hasn't kept up. In a world long accustomed to what-you-see-is-what-you-get displays, industrial knitting systems are souped-up versions of the old graph-paper code. After fashion designers use their own 3-D software to create new styles on-screen, someone still has to translate each garment design into a two-dimensional, stitch-by-stitch program. As digital knitting becomes increasingly powerful, that limitation frustrates apparel companies. They want to design in three dimensions.

The ribbed bunny embodies a solution to their problem.

Its knitted exterior was made on an industrial flatbed machine using an open-source visual programming system developed by Narayanan and her colleagues. (The stuffing—a "challenge," she says—was done by hand.) To create a knitted object, the designer can either feed an existing computer-aided-design pattern into the system, which will verify that it can in fact be machine knitted, or can use the system to create a new pattern on a 3-D model of the stitch mesh. Along with ribs and 3-D shapes, the system can generate lacy textures and multicolor designs.

Instead of forcing designers to think like knitting machines, in other words, the system lets them concentrate on what they want to make. "We're hoping that it's a more intuitive process for someone who's not a particular expert on knitting," says Narayanan. "You could work in the space of the 3-D object rather than working in the 2-D space." By hiding the code, the program paradoxically advances the legacy of Marx Ziegler, extending the power to create cloth beyond the current masters of the art.

Just as essential is a standard file format that works on any machine. The Carnegie Mellon lab has developed an open file format called Knitout. "It's just really simple," says McCann. "It just lists the operations you'd like the machine to do in the order you'd like it to do them." A design program like Narayanan's can generate a single Knitout file that, with the right translators, will produce the same knitted shape on a Shima Seiki machine, one from its German competitor Stoll, or models designed for small-scale production.

The Carnegie Mellon research is part of a multi-university initiative to significantly improve the digital design of fabric. Driven by the gaming and film industries, this endeavor began as work to create algorithms describing fabric and fiber characteristics so accurately that the virtual materials behave on-screen exactly as they would in real life. The goal is to encode the characteristics of every single type of fiber and thread into algorithms that will accurately mimic their behavior on-screen. What started with a desire for better animation now has apparel executives positively quivering with excitement. They imagine eliminating trial-and-error sampling, cutting lead times by three-quarters, and designing whole garments from the yarn characteristics up.

"The cost savings are absolutely mind-boggling, and the environmental impact," says Seiz, the knitting engineer. Echoing Ziegler, he sees shared prosperity coming from this new ability to encode and transmit patterns. "This is not something we're going to keep close," he says. "It's going to be shared with the public. All will be sharing in this bounty."[50]

Chapter Four

DYE

Everything visible is distinguished or made desirable by color.

—Jean-Baptiste Colbert, *General Instruction for
the Dyeing and Manufacture of Woolens*, 1671

SIX THOUSAND YEARS AGO, AROUND the same time as the founding of the
Mesopotamian city of Ur, someone on the coast of northern Peru tore apart
a piece of cotton cloth and left it at the ritual site known today as Huaca
Prieta, or the Black Mound. In a ceremony whose meaning is lost with the
ages, the cloth's owner bundled several fragments together, doused them
with saltwater, then smashed the decorated gourd from which it was poured.
As centuries passed, the area's dry climate preserved the cloth among hundreds of bits of textiles and gourds.

Inhabited for more than fourteen thousand years, Huaca Prieta and
its surroundings represent one of the world's earliest economically and
culturally complex settlements—possibly even *the* first. Here, at the rich
confluence of sea and river, wetland, desert, and plains, early people built
permanent villages, stored and traded food, and developed distinctive rituals and works of art. The artifacts they left behind contradict the long-held

archaeological assumption that agriculture and pottery necessarily go together. These ancient people grew crops but made no ceramics. In this area of abundant sea life and tropical fruits, including avocados, chilis, and cotton, people without pottery developed a complex way of life where gourds, nets, baskets, and cloth were essential tools.

The offerings they left at Huaca Prieta tell us that for these long-lost people, as for us today, textiles were more than functional artifacts. The cotton fragments aren't simply the tans and browns of the local cotton. They have blue stripes. Usefulness alone can't explain why someone would go to the trouble of making blue cloth.[1]

After a century and a half of textile colors scientifically concocted in laboratories, we privileged moderns take dyes for granted. But they're much trickier than you might think. "Any weed can be a dye," fifteenth-century Florentine dyers used to say. But that's only if you want yellows, browns, or grays—the colors yielded by the flavonoids and tannins common in shrubs and trees. Reds and blues are complicated and scarce, and greens all but impossible. Chlorophyll doesn't work as a dye.[2]

Only rarely can you simply put plant matter in hot water and dye fibers by steeping them in the solution. A few plants do yield tints that easily—onion skins, for instance—but most require additional chemical help, at least if you want the color to last more than a single washing.

Fortunately, dyeing offers clear results. Like metallurgy—and unlike medicine or magic—it either works or it doesn't. You can change variables and see what happens. Mistakes are unquestionably different from successes. Over time, techniques improve and people come to recognize patterns in the substances they use. Without knowing the underlying chemistry, early dyers learned to classify acids, bases, and salts by how they felt, smelled, tasted, and reacted. They knew that soft rainwater would behave differently from hard well water, with river water in between. They discovered that an iron dye pot would produce different shades than a copper or a ceramic one.

"The ancient dyer," writes chemist Zvi Koren, a specialist in the analysis of ancient colorants, "was an advanced empirical chemist." To produce long-lasting colors, he observes, the ancient dyer mastered the methods that are based on advanced chemical topics, such as, ionic, covalent, and intermolecular bonding, coordinate complexation, enzymatic hydrolysis, photochemical chromogenic precursor oxidation, anaerobic bacterial fermentative reduction, and redox reactions.

Not that the ancient dyer knew what those things were. Only since the eighteenth century have people had any idea what might be going on at the molecular level—and only since the nineteenth have we known that molecules even exist. "Of all the areas of human activities," writes French historian Dominique Cardon, in her monumental volume on the botany, chemistry, and history of natural dyes, "dyeing with plants is one of the best examples of the efficiency of the empirical acquisition of expertise."[3]

Dyes bear witness to the universal human quest to imbue artifacts with beauty and meaning—and to the chemical ingenuity and economic enterprise that desire calls forth. The history of dyes is the history of chemistry, revealing the power, and the limits, of trial-and-error experimentation without fundamental understanding.

𐐑𐐑𐐑𐐑

The blue in the Huaca Prieta cloth comes from indigo, one of the world's most popular botanical dyes. Its precursor compound *indican* occurs in a wide range of plants, growing in varied climates and soil conditions. *Woad*, the traditional blue of Europe, is related to cabbage. The South Asian plant *Indigofera tinctoria*, known in Europe as "true indigo," is a legume, as are the indigo species native to Africa and the Americas. Japanese indigo, or *tadeai*, belongs to the buckwheat family and is sometimes called "dyer's knotweed." These are only a few of the wildly different plants—each containing indican—that ancient people around the world realized could yield beautiful blue dye.[4]

Nowadays we call that plant-derived coloring "natural" to distinguish it from dyes formulated in chemical labs, including chemically identical synthetic indigo.[5] But producing indigo takes far more artifice and effort

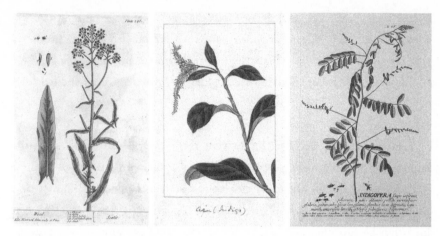

*People around the world have derived indigo dye from unrelated plants that all contain the same chemicals. From left: woad (*Isatis tinctorial*), the traditional blue of Europe; Japanese indigo (*Persicaria tinctorial; also known as* Polygonum tinctorium*); and* Indigofera tinctoria, *the South Asian plant better known simply as indigo. (New York Public Library Digital Collections; Library of Congress; Wellcome Collection)*

than the word *natural* implies. Its source may grow in the wild, but turning leaves into dyestuffs for making blue cloth requires considerable technological prowess. "In the modern world, we sometimes think of ancient people as primitive with a lack of understanding about the world," says archaeologist and textile specialist Jeffrey Splitstoser, who spearheaded the Huaca Prieta analysis. "But really, you had to be pretty smart to live back then."[6]

To produce indigo dye, you start by soaking the leaves in water. As their cells break down, they release indican, along with an enzyme. The enzyme catalyzes a reaction splitting the indican into sugar and a highly reactive molecule called *indoxyl*. The indoxyl bonds quickly with oxygen in the water, forming *indigotin*, the blue pigment otherwise known as indigo. Insoluble in water, the indigotin precipitates into a slurry at the bottom of the vat.

Now you have a stable pigment. That's great for paint or ink, but as long as the indigo won't dissolve, it can't adhere to cloth. To use it as dye, you have to change the pH of the water by adding a strongly alkaline (basic) substance, such as wood ashes. You can either start with alkaline water or separate the slurry from the original water and put it in a new vat, removing the spent leaves in the process.

COMPOUND	WHERE FOUND
Indican	Leaves
Indoxyl	Solution with enzymes from leaves
Leuco-indigo	Alkaline solution with low oxygen
Indigotin ("indigo")	Indoxyl exposed to oxygen

In a highly basic (alkaline) environment, the indigotin reacts to form a soluble compound known as *leuco-indigo*, sometimes called white indigo. "To dye with indigo, it is necessary seemingly to destroy it," explains Cardon, "although in fact it is transformed into a different substance that is soluble, but almost colourless, and can be absorbed by the fibres."[7]

Like indoxyl, leuco-indigo wants to bond with oxygen and turn into indigotin. To keep that from happening and maintain the solution, you have to reduce the water's oxygen levels. Dyers traditionally relied on bacteria either from the indigo leaves or from added foodstuffs such as dates, bran, or honey. They didn't know that bacteria or oxygen existed, let alone how the two interacted. They simply tried ingredients until they found ones that worked. Since the hidden chemistry was the same, many different additives could achieve the desired results.

"The substances found to be effective reducing and fermenting agents by different societies worldwide sound more like the ingredients for an extremely elaborate festive cake, since many of them are 'sweet,'" writes the indigo scholar Jenny Balfour-Paul.

They have included dates, grape and palm sugar, molasses, yeast, wine and rice spirit lees, local liquors, beer, rhubarb juice, figs, mulberry fruit, papaya, pineapple, ginger, honey, jaggery, henna leaves, wheat bran, flour, cooked glutinous rice and tapioca, madder, *Cassia tora* seeds, sesame oil, green bananas, sisal leaves, powdered betel nut, tamarind juice and, less appetizingly, putrefying meat.

As chemical knowledge progressed in the eighteenth century, dyers began removing oxygen from the dye vat by introducing iron compounds that would bond with it and precipitate out as rust. Regardless of ingredients, indigo dyeing requires a highly basic solution with low levels of free oxygen.[8]

As the leuco-indigo dissolves, the solution below the surface turns a weird antifreeze-like yellow-green. At the top of the vat, iridescent purply-blue bubbles form—a sign of indoxyl bonding with oxygen in the air. You can now immerse your fiber, yarn, or fabric. Dissolved in the water, the leuco-indigo penetrates the microscopic nooks and crannies of the fibers, turning them a greenish shade. When you lift the material out of the vat, exposing it to the air, the leuco-indigo bonds with oxygen to form indigotin. The fibers turn blue, as if by magic. Repeated dipping makes the color darker, adding layers of indigotin molecules.

Dyers don't have to use the indigo immediately. The slurry may be preserved as a damp paste or rolled into balls with the leaves for later reconstitution, the common historical practice with both woad in Europe and *tadeai* in Japan. Or it may be dried into cakes, which are light, durable, and easily transported—ideal for long-distance trade. Beginning in the sixteenth

Indigo processing, as depicted by Pierre Pomet's L'Histoire Générale des Drogues *in 1694. It isn't clear whether the scene is supposed to be set in India or the West Indies. (Internet Archive)*

century, European merchants brought indigo cakes from India, giving the dye its name and gradually replacing the weaker pigment concentrations in native woad.[9]

Seeing only the beautiful results, it's easy to overlook a major drawback to all this indigo chemistry: it stinks. "It took me awhile to understand," writes a first-time indigo dyer, "that the ripe, sweaty outhouse scent suffusing the air was coming from the giant pot of dye, not a backed up toilet." So smelly was fermenting woad that Elizabeth I banned it from an eight-mile radius of any of her palaces. Pollution didn't start with the Industrial Revolution.

At an indigo-dyeing workshop in Los Angeles, textile designer Graham Keegan passes around a jar of concentrated slurry, encouraging everyone to sniff it. One by one, we gasp. "Whew," says a student. "It lingers." Playing the connoisseur, Keegan pronounces this year's vintage subtler than last year's. "It's got a lot of rotten, some fecal notes, a lot of urine in there," he says. "It's got it all."[10]

The result of this odoriferous chemical magic is an unusually colorfast dye. The blue doesn't dissolve in the wash or fade in the sunlight. "The only original colour of the Bayeux *tapestry* that remains true," Balfour-Paul writes, "is the indigo blue of its woad-dyed wools." (Italics indicate a word in the glossary.) And unlike most plant dyes, indigo adheres easily to cellulose fibers such as cotton and linen. The Huaca Prieta cloth was cotton, and about 4,400 years ago, Egyptians wrapped a mummy in otherwise undyed linen with fine indigo stripes. Indigo does rub away with wear, as we see with indigo-dyed blue jeans. But it can stay true for thousands of years, testifying to the ingenuity of people from civilizations long past.[11]

"Despite all the experimentation that must have taken place before full mastery of dyeing with indigo was achieved," writes Cardon, "archaeological discoveries show that this skill had already been acquired in prehistoric times by several geographically widely separated civilizations."[12]

The idea seems to have originated with the careful observation of chance occurrences. Someone saw indigo leaves that turned blue in an early frost or noticed the striking colors produced when a summer storm ripped leaves off an indigo-bearing plant and blew them into the wet remains of a wood fire.

"The plant appears blue whenever it gets damaged," says Keegan, who uses natural dyes in his design work. "If the indigo was in a rainstorm and it fell in a puddle, that puddle would turn bluish green. And the same skin that's on top of the vat, that purple-coppery, would be sitting right on top of the puddle."[13]

To test this assertion, I take Keegan up on his offer of fresh indigo leaves, dividing them into two bunches. One I put in plain tap water, the other in water mixed with ashes from burning some bamboo skewers. Nothing happens when they sit on my kitchen counter overnight. The problem, I surmise, is the temperature. Both the water and the house are too cool.

I transfer the leaves and ashes to glasses of warm water and move them to a small bathroom I can keep warm by running a heater. Sure enough, the next day tiny particles of blue appear on the surface of the plain water. The ashy water is a copper color. After two days, it has a slight green tinge. When I immerse a bit of white cloth in it a few times, the cloth turns a pale blue-gray. It isn't the striking transformation produced by Keegan's concentrated dye vat, but it proves the point. I've re-created the primordial puddles.

𝕂𝕂𝕂𝕂

A Phoenician legend recounts the origins of Tyrian purple dye. One day, it says, the god Melqart, the lord patron of Tyre, was walking on the beach with his mistress and his pet dog. The dog fished a sea snail out of the water and began eating it. As he bit down, the dog's mouth turned purple. The transformation inspired Melqart to use the snails to dye a tunic for his beloved and to bestow upon Tyre the dye secrets that made the city rich.[14]

The Phoenicians who told this tale were the great navigators and merchants of the ancient Mediterranean, sailing from their home ports in what is now Lebanon to destinations as far away as Spain's Atlantic coast. Along with cargos of cedar wood, glass vessels, and Tyrian purple, they brought with them the script that evolved into most of the world's alphabets.

As the legend suggests, it's not hard to imagine how people discovered the source of the most famous color of antiquity. Not plants but sea creatures furnished the precious purples of Persian royal robes, Hebrew priestly raiment, and imperial Roman togas. Throughout the Mediterranean, mounds of discarded shells testify to what was once a major industry, consuming thousands of snails for a single dye bath. Ancient people expended enormous effort and ingenuity to obtain shellfish purple, demonstrating both the love of color for its own sake and the lust for luxuries signifying social status.

Dyers extracted colorants from three different species of snails, using their dyes alone or in combination to produce different shades. The spiny dye murex, *Bolinus brandaris*, and the red-mouthed rock snail, *Stramonita haemastoma*, yielded a red-violet. The banded dye murex, *Hexaplex trunculus*, was more versatile. Its fluid could turn blue, blue-violet, or red-violet. Modern scientists understand the chemistry. The blue is our old friend indigotin, the blue-violet has an added atom of bromine, and the red-violet contains two bromine atoms. But researchers still debate how ancient dyers could predict which colors they'd get. Did the results vary by subspecies, sex, or conditions?[15]

We know much about ancient purple dyeing from the Roman author Pliny the Elder, who described the process in his encyclopedic *Natural History*, published between 77 and 79 CE. Unfortunately, his seemingly detailed accounts lost crucial information with the passage of time. Within the same species, for instance, he distinguishes between snails that feed on seaweed, on rotting slime, and on mud, as well as a variety collected from reefs and another named for a pebble. To ancient dyers, these were meaningful categories, offering clues to the shades expected from each. Today, they're mysteries.

Besides, "Pliny was a reporter, not a dyer; he told us what he saw without really understanding the procedure," observes an artist who re-creates ancient dyes and pigments. Using chemical principles, modern experimenters have tried to figure out how the process might have worked.[16]

The murex, Pliny explained,

has the famous flower of purple, sought after for dyeing robes, in the middle of its throat: here there is a white vein of very scanty fluid from which that precious dye, suffused with a dark rose color, is drained, but the rest of the body produces nothing. People strive to catch this fish alive, because it discharges this juice with its life.

The living mollusk contains colorless forms of indoxyl. When the snail dies, it releases an enzyme that leads these compounds to bond with oxygen and produce the colored fluid. (Depending on the specific compound, the process may require sunlight as well.) Rather than risk letting the color escape into the sea, ancient harvesters collected live snails and preserved them in tanks. Archaeological dye sites sometimes contain murex shells with distinctive holes bored in them—the result of storing large numbers of mollusks in close quarters and not adequately feeding them. Without other prey, the snails cannibalize each other, using an acid-secreting organ to punch a hole in another's shell so they can extract the flesh.[17]

Once dyers had enough snails, they opened the shells and cut out the glands containing the pigment; those too small for this surgery, they simply crushed. They put the glands, secretions, and crushed snails into a vat of water kept warm, but not boiling. To this stew, they added what Pliny calls *salem*, Latin for *salt*. They let the concoction steep for three days, producing a concentrated solution. This they transferred to a metal cauldron, Pliny reports, then added water, skimmed off the snail residue, and let the solution cook for another nine days until a test fleece proved it ready for dyeing.

Pliny's mention of *salem* puzzles modern chemists, because ordinary table salt, the stable compound sodium chloride, wouldn't aid the dyeing process. "It seems surprising," writes Cardon, "that no ingredient other than mollusc and salt is mentioned." Surely purple dyers must have added something to produce an alkaline bath, similar to the one used for indigo. Cardon observes that archaeologists often find purple dye workshops near lime or pottery kilns, both of which could supply dyers with alkaline ash.[18] It's a reasonable argument, but, as we'll see, Pliny may have gotten the ingredients right after all.

Accustomed to the brilliant tones of synthetic dyes, we tend to imagine ancient purple as a bright color. But the Tyrian purple worth its weight in silver wasn't the Technicolor hue bedecking Rex Harrison as Julius Caesar in 1963's *Cleopatra*.[19] Pliny described it as "the color of coagulated blood: dark when observed from the front, with bright reflections when seen from an angle." In later Latin it was called *blatta*, from the word for "clot." The most valuable of ancient dyes was not, by today's standards, an especially attractive color.

It also stank—and not just during the dyeing process. Pliny's younger contemporary, the satirical poet Martial, listed "a fleece twice drenched in Tyrian dye" in a litany of terrible smells and joked that a rich woman dressed in purple because of its odor, hinting that it masked her own. "What is the cause of the prices paid for purple-shells," clucked a disapproving Pliny, "which have an unhealthy odor when used for dye and a gloomy tinge in their radiance resembling an angry sea?"

For buyers, the answer was social status. Few could afford Tyrian purple, so it marked its owner as special. Writing in the early sixth century CE, the Roman statesman Cassiodorus termed the color "an ensanguined blackness which distinguishes the wearer from all others." The rare hue stood out in a crowd. Martial wrote in his epigram "On the Stolen Cloak of Crispinus":

Crispinus does not know to whom he gave his Tyrian mantle, when he changed his dress at the bath, and put on his toga. Whoever you are that have it, restore to his shoulders, I pray you, their honours; it is not Crispinus, but his cloak, that makes this request. It is not for every one to wear garments steeped in purple dye; that colour is suited only to opulence. If booty and the vicious craving after dishonourable gain possess you, take the toga, for that will be less likely to betray you.

Even the purple's notorious stench conveyed prestige, because it proved the shade was the real thing, not an imitation fashioned from cheaper plant dyes.[20]

For sellers, the high cost reflected how laborious and disgusting the dye was to produce—as archaeologist Deborah Ruscillo learned in the summer of 2001. A specialist in analyzing animal remains, Ruscillo was intrigued by the enormous piles of murex shells she found at archaeological sites and wondered just how many snails ancient dyes required. With a graduate assistant, she decided to follow Pliny's instructions and find out.

To start, Ruscillo baited traps in a bay off Crete, where *H. trunculus* feast on fish discarded by local fishermen. She quickly discovered that the traps filled with water, making them heavy to lift. They also captured unwanted cargo. "Eels and ground feeders like scorpion fish found their way into the traps, potential hazards for any diver raising baskets or pots," she writes. The good news was that the bait enticed more murex to the surrounding seabed. Along with those in the traps, Ruscillo and her assistant could each gather another hundred snails an hour. Archaeological remains confirm that ancient murex harvesters must have employed both traps and hand collection; shell mounds contain tiny snails that hand gatherers would have left behind and already-dead snails that couldn't have gotten into traps.

With more than eight hundred snails in a bucket of seawater, the two researchers moved to a site "well away from the modern village." Ruscillo knew that ancient dye works were located far from settlements, and she soon experienced the reason firsthand.

But first they had to get the shells open. "Good lord, that's impossible," she recalls in an interview. "It's rock." Taking a hint from the holes in ancient remains, she developed a two-stage technique: whacking the main whorl with a brass awl to create a small hole, then prying open the shell.

Next came the gross part. They cut out the glands and discarded the rest, generating piles of broken shells that closely resembled the ones archaeologists find at ancient sites—with one important difference. Unlike the ancient shells, these contained decaying meat. Immediately, Ruscillo says, "you are surrounded by flies—big horseflies that are biting you—and wasps."

They put the glands into a covered aluminum pot of water, which became an increasingly vibrant purple as more were added. Even the tight lid couldn't stop flies from laying eggs in the slimy mixture. "The big flies were sitting on the rims, laying these larvae and shoving them under the lid with their legs," Ruscillo remembers. "It was remarkable." To kill the resulting maggots without ruining the dye, she had to heat the solution to just below boiling.

Whereas ancient dyers used vats of a hundred liters or more, Ruscillo conducted her experiments with small pots, each containing about twenty ounces (590 milliliters), enough to dye a sample swatch measuring six inches by eight inches. Even at this small scale, she encountered the dye's legendary stench. "The great number of dye-works makes the city unpleasant to live in," the Greek geographer Strabo wrote of Tyre, "yet it makes the city rich through the superior skill of its inhabitants."[21]

"Unpleasant" is a serious understatement. "The workmen eating fifty meters away were complaining about how stinky it was," says Ruscillo. To endure the stench, she and her colleague had to wear masks. In an ancient dye center, the overwhelming odor would have been multiplied many thousandfold. The work also stained their hands a purple no amount of washing could remove. This awful job, Ruscillo concluded, must have been done by slaves.

Ruscillo tested four different fabrics—wool, cotton, nubby raw silk (silk noil), and smooth satin silk—and determined that wool and satin silk took the color well. For the dye baths, she used ingredients ancient textile specialists suggested: seawater alone, fresh water alone, seawater and urine, seawater and the mineral salt *alum*, and seawater and vinegar.

She quickly realized that Pliny himself had never dyed with purple. When the experimenters followed his instructions for two stages of heating, one for three days and a second for nine, they got an unimpressive gray with a slight purple cast. Ruscillo decided to try skipping the nine-day step. Instead, she simply steeped each mixture for three days at 80 degrees Celsius, then filtered out the snail remains and immersed the fabric samples. She left the fabric to slowly cool in the dye. As a sixth test, she used seawater without any steeping.

Depending on how many murex were used and how long the fabric soaked, the dye produced colors ranging from pale pink to a black-purple. "All of them were very beautiful colors," she says. The silk's vibrant hues were more to modern tastes, while wool achieved the darker shades preferred by the ancients. It also sucked up much more dye. "The wool absorbed the dye immediately like a sponge," Ruscillo writes, "and maintained the deep colour even after rinsing." The fabrics also held the stench. Nearly two decades later, they still smell—despite washing in Tide.

Surprisingly, Ruscillo got essentially the same results from every solution. The urine made the purple more vibrant, but in general, she recalls, "With the same amount of time, the same concentration, and the same amount of water," added ingredients "did not seem to make a difference in the hue." (Seawater did make the dye more colorfast than fresh water.)

Unencumbered by chemical theory, Pliny might have been right about the ingredients after all: salt, to preserve the snail flesh from rotting, and seawater. Maybe dyers didn't create an alkaline bath. Maybe seawater alone did the trick. It is somewhat alkaline, with a pH around 8.3, with 7 being neutral.[22]

The biggest revelation came when, almost as a lark, Ruscillo tested the batch using seawater without steeping the dye. After immersing the samples for just ten minutes, she writes,

> I watched as the white, slimy swatches dried to beautiful blue hues. This experiment had recreated the "Biblical Blue" or *tekhelet*, as it is known, sacred in antiquity as well as the present, particularly in the Jewish religion. It is known that this sacred blue was produced from marine snails. The *talit*, or ritual prayer cloth worn by men during morning prayer or the wedding ceremony, was traditionally to have a blue fringe dyed from *Murex*.

This rich blue, akin to the color of indigo-dyed light denim, was a wholly unexpected result.

Ruscillo's experiments demonstrate just why ancient purples were costly and rare. "The number of human hours used to produce one garment is substantial, not considering the arduous tasks *Murex* dye production involved," she writes. To dye trim or a lightweight garment, she found, would require a few hundred snails. A large wool cloak like the one swiped from Crispinus would consume thousands, representing hundreds of hours for harvesting alone.

Ruscillo's research is particularly noteworthy because she didn't start with chemical theory about what ancient dyers must have done. Instead, she tested common ingredients and observed what happened. She recapitulated the kind of trial-and-error learning that ancient dyers relied upon, but with modern science's experimental rigor. The process was systematic and scientific but entirely empirical. And in this case, it demonstrated results that theory-based prediction missed.[23]

For most of history, dyeing was more like cooking than chemistry. It involved chemical reactions, but the dyer didn't necessarily understand them. Different people followed different recipes, and many crucial techniques weren't written down. They passed from master to apprentice through hands-on practice. With no standard measuring instruments for temperature or pH, good results depended on carefully observing color, smell, taste, texture, even sound. There might be more than one way to achieve the same hue. Most likely, therefore, some purple dyers used alkaline additives and some did not, some used seawater and others did not, and some used urine or vinegar or some other secret ingredient. Some customary ingredients made a difference in the results, and others were actually unnecessary. Results varied with the dyer, and reputation mattered.

In the frescoes he painted on the walls of Florentine churches, Domenico Ghirlandaio included many portraits of eminent citizens, depicting them as witnesses to holy events. From these paintings, we know the faces of fifteenth-century humanists, bankers, and their families. We also know they loved wearing red.

Nearly every man not in religious habit has on a red cloak, often with a matching hat. Women appear with red sleeves or pink dresses. Bed covers and curtains feature shades of red. In the self-portrait he tucked into his *Adoration of the Magi*, Ghirlandaio himself wears scarlet. So do the grandfather with the wart-covered nose and his towheaded grandson in the famous painting now in the Louvre. By the Renaissance, red had replaced royal purple as the symbol of wealth and power.

So it's not surprising that when a Venetian named Gioanventura Rosetti published the first professional dyeing manual in 1548, the largest number of recipes, thirty-five, were formulas for red. Black followed with twenty-one.

Titled *The Plictho*, Rosetti's work represents sixteen years of effort, much of that time undoubtedly spent prying trade secrets from reluctant artisans. The author was not himself a dyer but, rather, someone who believed in disseminating technical knowledge; he later wrote a similar book on perfumes, cosmetics, and soaps. Describing *The Plictho* as "a work of Charity that I bequeath for the public benefit," Rosetti complained that dyeing know-how "has been imprisoned for a great number of years in the tyrannical hands of those who kept it hidden."

A page from the 1560 edition of Gioanventura Rosetti's Plictho *(Getty Research Institute collection via Internet Archive)*

Like Marx Ziegler's weaving manual a century later, *The Plictho* represents the first stage in a knowledge revolution: recording and publicizing state-of-the-art technology and practice. Rosetti didn't seek to analyze or improve the formulas, only to make the information available so that others could learn from them.

Dyeing, he wrote, "is an ingenious art and fit for acute intellects." His recipes illustrate how far dyeing practice could progress on a purely empirical basis, without good (or any) chemical theory. They also capture dyeing on the cusp of a new age, before colorants from the Americas were fully adopted in Europe.

Consider a simple recipe titled "To dye wool or cloths in red."

For each pound of wool take 4 ounces of roche alum and make it boil one hour. Wash it very well in clear water. Then after it is well washed, take for each pound of wool 4 ounces of madder and make it boil in clear water.

Throw in the madder when it is about to boil, then the wool, and let it boil for half an hour, stirring constantly. On washing it becomes well dyed, that is, red.[24]

This formula employs one of history's most important dyestuffs: the roots of *Rubia tinctorum*, better known as dyer's *madder*. This widely cultivated species, Cardon writes, "takes pride of place in the history of dyeing" because of "the astonishing variety of colours that could be obtained from it—either alone or in combination with other dyestuffs." Textile fragments excavated from Masada, the desert palace fortress famous for the mass suicide of Jewish rebels in 73 CE, include bright red, salmon pink, deep burgundy, purple, purplish black, and reddish brown—all dyed with madder.[25]

Madder's versatility derives from the chemistry exploited by two types of traditional dyers' know-how. The first is botanical. The root contains two different color-producing chemicals: *alizarin*, which yields an orange-red, and *purpurin*, which creates a purplish hue. The proportions vary depending on the subspecies of plant, the soil conditions, and the root's age when harvested. By taking advantage of this variation, dyers could produce a range of shades.

They also knew that different additives could alter the color. To give madder a bluish tone and soften hard water, Renaissance dyers relied on *bran water*, an acid made by soaking bran for several days. (Properly prepared bran water "smells like vomit," observes a dyer who re-creates Renaissance techniques using sources including *The Plictho*.) *White tartar*, a sediment produced in wine fermentation, would give the red an orange cast.[26]

The most significant additives were *mordants*, from the Latin word *mordere*, meaning "to bite." These are chemicals, usually metal salts, that cause madder and most other so-called natural dyes to securely attach to fibers. That's what the first step in the *Plictho* recipe is about. Before dyeing, the material is soaked in the mordant, in this case alum. The alum bonds with the fibers and, when the wool is immersed in the dye pot, the mordant provides a bridge to bond with and fix the dye. Here, again, we see the application of trial-and-error empiricism. Even today chemists debate exactly how fiber molecules, mordants, and dyes interact.[27]

Different mordants lead to different final shades. Iron compounds, for instance, dull and darken the color. The reddish-brown Masada textiles used iron with madder, and as early as the fourteenth century BCE, Egyptians employed iron mordants with tannins from plants to create browns and blacks. Some of Rosetti's formulas for black work the same way, while others rely on gallnuts, whose tannins serve as mordants while darkening colors.[28]

The most important mordant is the one in the *Plictho* recipe: *alum*, a potassium (or sometimes ammonium) aluminum sulfate that not only fixes but brightens colors. Fairly pure crystals of alum occur naturally in deserts and volcanic areas, explaining the prehistoric origins of its use. By classical antiquity, however, people had learned how to extract large quantities of usable alum from the mineral alunite, which is found in volcanic areas. To get alum from alunite, you first heat the rocks in a kiln, then repeatedly pour water over them until they form a paste. You boil the paste, separating the insoluble compounds, and decant the solution. It will then crystallize into purified alum.

In Rosetti's day, alum mining, production, and trade were big businesses—the first international chemical industry. In 1437, for instance, Florentine traders made a five-year deal to buy about two million pounds of powdered alum from Byzantine sources. "Alum is no less necessary to dyers of wool and woolen-cloth than bread is to humankind," declared a sixteenth-century writer.[29]

One of Rosetti's recipes promises "orange loaded with color." It uses twenty pounds of alum along with fustet, a yellow from the European smoke tree, and three different reds: madder, *brazilwood*, and *grana*, an expensive scarlet derived by grinding up thousands of tiny insects.[30] Loaded with color indeed.

Here we see European dyeing on the cusp of change. Brazilwood, which comes from the red heartwood of certain trees, had once been a rare dye derived from Asian sappanwood imported by Venetian traders. By Rosetti's day, it was plentiful, thanks to abundant supplies in the American tropics. In just one month of 1529, notes art historian Mari-Tere Álvarez, Spain imported a stupendous six thousand tons from its New World territories.

The dense wood from which Brazil got its name was both better than the Asian alternative and much cheaper—so cheap, in fact, that Álvarez complains that scholars don't take it seriously enough. To art historians, she says, it's the "Costco of pigments." As a dye, brazilwood does leave much to be desired because it quickly fades in the light to a dull brick color. But it can add depth to longer-lasting colorants. Rosetti's recipes use it primarily as a supplement to madder or *grana*.[31]

"Since many of these formulas appear to be those of skilled commercial dyers," write *The Plictho*'s translators, "we can be reasonably sure that by 1540 brazil had become an important red dyestuff, along with kermes [*grana*] and madder. This extensive use in commercial dyeing could only have come about because of its new availability at a very low price, for its properties were exceedingly poor."[32]

Fig. 290. — Cactus-Nopal portant des cochenilles.

Cochineal insects growing on the nopal cactus, with a much-enlarged depiction of the insect's appearance (Internet Archive)

The New World supplied more than bargain-basement reds. In 1500, Europe's best and most valuable source of red dye was *kermes*, made from a tiny insect that lives on European oak trees. Fifty years later, it was Mexican *cochineal*, derived from a similar parasitic insect that grows on the nopal, or prickly pear, cactus. Because the tiny dried bug carcasses look like plant or mineral grains, European dyers used *grana* to refer to both dyestuffs. The translators assume that when Rosetti uses the word *grana* he means kermes, but we can't be completely sure. As Rosetti was conducting his research, European dyers were switching raw materials.[33]

Containing ten times as much colorant as kermes, cochineal was one of the New World's greatest gifts—a tribute to Mexico's native farmers. While kermes grew wild, indigenous people had cultivated cochineal for centuries. Like the Chinese who domesticated silkworms, they devoted careful attention to both the insect and its host plant. By the time the Spanish arrived, their selective breeding had produced "the closest thing Europe had ever seen to a perfect red," writes Amy Butler Greenfield in her history of the dye.[34]

As a dye, cochineal was brighter, more colorfast, and easier to use than its competitor. As a trade good, it was ideal: lightweight and costly. Rival Tlaxcaltecan and Aztec merchants had sold it throughout the region before the Spanish conquest. By the mid-sixteenth century, it was one of New Spain's most valuable exports.

Like the Aztec rulers before them, Spanish authorities collected cochineal as tribute. But taxes alone weren't enough to satisfy European demand for the dye. Cochineal farming soon became a lucrative commercial venture— so much so that it upset the social status quo.

In 1553, the ruling council of Tlaxcala fretted that peasants were making too much money from cochineal, leading them to substitute the cash crop for subsistence farming; instead of growing their own food, they bought it in the marketplace, pushing up prices. Council members, who had been the precolonial elite as well, deplored the conspicuous consumption of the cochineal *nouveaux riches*. "Both the cactus owners and the cochineal dealers, some of them, sleep on cotton mats, and their wives wear great skirts, and they have much money, cacao, and clothing," they complained. "The wealth they have only makes them proud and swaggering. For before cochineal was known and everyone planted cochineal cactus, it was not this way."[35] Cochineal had been known for centuries, of course. What was new was a large overseas market.

Cochineal exports grew steadily over time, averaging between 125 tons and 150 tons a year by the end of the sixteenth century. Annual fluctuations could be extreme, however. In 1591 shipments totaled 175 tons, dipping to 163 in 1594 and plummeting to half that in 1598. Every year, European textile producers anxiously awaited news of how much the annual flotilla from New Spain might bring—and how much the year's cochineal would cost.

Any hint was potentially valuable business intelligence. "From every trading center in Europe," writes a historian,

> came factual reports, estimates, and guesses as to the amount of *grana* that would be available for the year's trading. The latest news from Brussels concerning the "cochineal fleet" was relayed from Rome in 1565; an Antwerp dispatch estimated total receipts of 1580; a letter from Mexico reported the sailing of one ship with a cochineal cargo in 1586.

By 1600, the New World insect was an essential dyestuff. Venetian kermes merchants lost their grip on the high-end market for red dyes, replaced by Spanish imports of cochineal flowing through Amsterdam and Antwerp. From 1589 to 1642, the Amsterdam price of cochineal quadrupled.[36]

Spain jealously guarded its monopoly by forbidding foreign ships to carry cochineal. Pirates and smugglers worked to break that stranglehold. With Elizabethan England in a cold (and occasionally hot) war with Spain, English privateers made cochineal-bearing ships their targets. The biggest haul was that of Elizabeth's favorite Robert Devereux, second Earl of Essex, who brought home more than twenty-seven tons of cochineal when he captured three Spanish ships in 1597. In the portrait painted shortly thereafter, Essex posed in a rich scarlet robe, no doubt dyed with New World red.[37]

ꗃꗃꗃꗃ

The Americas gave European dyers new color sources. Trade with India gave them competition and inspiration.

When the Portuguese reached India in the sixteenth century, they brought back cloth unlike anything in Europe: lightweight cottons decorated in rich colors that withstood washing. Finely spun Indian cotton was itself a wonder—soft, cool, and washable, a miraculous alternative to scratchy linen, difficult-to-clean wools, and costly silks.

Then there were the colors. As a cellulose fiber, cotton resists most dye. Yet Indians had mastered techniques for a rainbow of hues: reds, blues, pinks, purples, browns, blacks, yellows, and greens. Unlike European textiles, whose designs were woven in or embroidered, these Indian cottons featured

An eighteenth-century painted palampore, or bed cover, made in India for the Sri Lankan market. These much-prized textiles were also used as wall hangings and tablecloths. (Metropolitan Museum of Art)

multicolored painted or printed motifs. The textiles that came to be known as *chintz, calicoes,* and *indiennes* were a revelation.

"In some things the Artists of *India* out-do all the ingenuity of *Europe, viz.,* the painting of Chites or Callicoes, which in Europe cannot be parallel'd, either in their brightness of the Colours or in their continuance upon the Cloath," wrote the British East India Company chaplain John Ovington in a 1689 account of his travels in western India. By then, the East India Company was bringing home more than a million pieces of calico a year, about two-thirds of its total trade.

Europeans couldn't get enough of the new cloth. To compete with Asian prints, domestic dyers needed to up their game, improving existing processes and mastering new techniques.[38]

It was in this context that Jean-Baptiste Colbert, Louis XIV's powerful finance minister and the father of French *dirigiste* economic planning, argued for tighter controls over the dye industry. "If the manufacture of silk, wool, and thread is that which serves to sustain and make commerce pay," he wrote to the king in 1671,

dyeing, which gives them a beautiful variety of colors such as are found in nature, is the soul without which the body would have but little life. . . .

Not only is it necessary that colors be beautiful to increase the commerce of cloth, but they must be of good quality so that they may last as long as the fabrics to which they are applied.

Colbert moved to publicize effective formulas, invest in research, and enforce uniform standards. His seemingly sensible demand that dyers adhere to the best-known practices, however, contained a contradiction. The policy simultaneously banned new methods and rewarded dyers who came up with improvements. "The fruits of experimentation were rewarded," observes a historian of science, "while the experiment itself was illegal."[39]

Indian dyers had developed their methods through the kind of trial and error Colbert's program frowned upon. Over the centuries, they had unquestionably made progress. But without a scientific foundation, they didn't know where further improvements might lie or what current steps or

Dyeing at the Gobelins dye house, as depicted in the eighteenth-century Encyclopédie. *Originally owned by the Gobelin family, the dye works and surrounding grounds were purchased in 1662 by the French state under Colbert to serve as suppliers to the court and a research center for improving dye formulas and techniques. (Wellcome Collection)*

ingredients might be extraneous. The European calico craze, by contrast, coincided with the development of chemistry as a science.

In 1737, the French government began appointing a leading chemist as the inspector of dye works, a far more prestigious job than the title suggests. The post was "the best place for science," a disappointed aspirant said.[40] It paid well and supported advanced chemical research. These scientists conducted experiments, gave lectures, and published books probing the mysteries of why certain substances gave fibers color, why some dyes lasted and others faded, and how to determine which was which. Was dyeing a chemical or physical process? How did it relate to Isaac Newton's theories of optics? Did dye coat fibers like a paint or glaze, or was something else going on? Succeeding inspectors offered their own theories and respectfully poked holes in their predecessors'.

In these early years of chemistry, dye processes were more likely to inspire chemical experiments than science was to generate new dyes. Working in a dye house was a way to stay on the cutting edge of scientific thought. That's why, after a taste of chemistry, twenty-year-old Jean-Michel Haussmann abandoned his pharmacy studies and joined his older brother Jean at a German textile printing company. While Jean concentrated on business affairs, Jean-Michel mastered dyes. In 1774, the brothers set up their own textile printing business in Rouen; the following year, they moved it across France, to Logelbach in Alsace.

Right away Jean-Michel faced a test of his chemical skills. The same dye processes that had produced bright colors at the old factory turned out dull hues in the new one. Instead of a brilliant scarlet, madder-dyed cotton came out a drab brownish red—not at all what customers wanted. Jean-Michel conducted experiments to find the culprit, eventually determining that the critical variable was the local water. It was too soft. The limestone in Rouen's water, he concluded, had removed a substance that dulled the red. By adding chalk to Logelbach's water, he achieved the same brilliant hues.

"Haussmann had the great merit of applying science to industry," concludes a local chronicle. "His knowledge of chemistry allowed him to imitate the beautiful colors that made *chinoise* cottons so desirable."[41] But it was the *practice* of chemistry, not its theory that made the difference. A young chemist like Haussmann knew how to conduct systematic experiments, controlling variables that might affect outcomes. Chemists still had only the haziest idea what was actually happening.

Calcium was not identified as an element until 1804, and the very idea of elements and compounds was itself new. Chemists still explained dye as a physical interaction of colorant particles and fiber pores; as a chemical change created by the dye's interaction with phlogiston, a substance believed to be contained in all flammable matter; or as some combination of the two.

While Jean-Michel was investigating dye results in Logelbach, his countryman Antoine Lavoisier was doing the experiments that would revolutionize chemistry. Combustion, he determined, had nothing to do with phlogiston. It occurred when a substance combined with the newly discovered breathable gas that the English scientist Joseph Priestley called "pure air" and Lavoisier named *oxygène*.

In 1789, Lavoisier published his groundbreaking book *Traité élémentaire de Chimie*, known in English as *Elements of Chemistry*. It laid out the concepts of elements, compounds, and oxidation, along with the system for naming chemical compounds used to this day. "As a textbook," observes the American Chemical Society, "the *Traité* incorporated the foundations of modern chemistry."

> It spelled out the influence of heat on chemical reactions, the nature of gases, the reactions of acids and bases to form salts, and the apparatus used to perform chemical experiments. For the first time, the Law of the Conservation of Mass was defined, with Lavoisier asserting that ". . . in every operation an equal quantity of matter exists both before and after the operation." Perhaps the most striking feature of the *Traité* was its "Table of Simple Substances," the first modern listing of the then-known elements.[42]

Among Lavoisier's enthusiastic early supporters was Claude Louis Berthollet, who held the dye inspector post. In 1791, Berthollet published his own landmark book applying the new chemistry to dyes. "He analyzed dyeing in the same way which he would approach any other chemical question and related the dye's chemical composition to the properties of the dye substance," writes textile historian Hanna Martinsen. "This approach represented Berthollet's view and *changed textile dyeing from a craft, based on traditional recipes and accidental improvements, to a contemporary technology based on scientific knowledge and systemic improvements.*" (Emphasis in the original.)[43]

That was the ideal, anyway. In fact, like manuals going back to *The Plictho*, Berthollet's book included many recipes without theoretical bases. Chemistry was still in its earliest days as a science after all, and much remained unknown. As a by-product of his chemical research on oxygen, for instance, Berthollet himself developed chlorine bleach—a major advance over the months-long process of repeatedly treating cloth with lye (a base) and buttermilk (an acid) and spreading it out on acres of grass. Yet he never understood that chlorine was not an oxygen compound but an element in its own right.[44]

Despite the many unknowns, the new chemistry gave dyers explanations for phenomena that had long been puzzling. Finally, they understood the transformation of indigo from blue to almost clear to blue again—and why blue foam covers the top of an indigo vat. "It appears that indigo passes . . . through different degrees of deoxydation," wrote Berthollet,

> so that its solution assumes different shades. In the most advanced state its solution is colourless; with less oxydation it passes to a yellow, and finally to a greenish hue.
>
> While indigo is in solution, the portion of it in contact with the air absorbs oxygen, which combines with the indigo, and regenerates it, saturating, at the same time, the substance which tended to seize it, so that the surface becomes blue. Thence the froth, green at first, and then blue, is called *fleurée* which is formed in well constituted vats, when they are agitated.[45]

With no idea of the structure of molecules, Berthollet couldn't fully account for the complex transformations of the indigo process. But Lavoisier's principles at least put him on the right track. Instead of focusing on why dyes create specific colors—a question that would require quantum physics to answer—dye chemists began to emphasize the reactions taking place. As

Newton gave way to Lavoisier and phlogiston was succeeded by molecular models, chemistry gained power traditional dyers could only have dreamed of. Within a century, laboratories would be creating so many new dyes that simply naming them would prove a challenge.

꧁꧂꧁꧂

In silhouette and provenance the museum's silk taffeta dress is nothing special. Suitable for paying afternoon calls, it has the high neckline, bell skirt, and slim waistline characteristic of 1860 fashions. Buttons down the front indicate an owner who dressed without a ladies' maid, while close inspection reveals sweat stains in the armpits. The seamstress nicely employed the dress fabric as piping and trim, but the design is hardly cutting-edge, nor is the construction *haute couture*. Whoever made it used a sewing machine.

Yet this undistinguished garment is not in a history museum but in the Museum of the Fashion Institute of Technology, a New York institution dedicated to clothing that "moves fashion forward." When I saw it in an exhibit on the history of color, I immediately knew why. Nothing that came before matched the intensity of its black and purple stripes. Here, manifest in a single garment, was the earth-shaking arrival of synthetic dyes. Once you've seen such deep blacks and vivid purples, or hot pinks and malachite greens, your visual expectations are never the same.

Influenced by black-and-white illustrations, delicately tinted engravings, and Queen Victoria's widow's black, we often imagine nineteenth-century European ladies as somberly dressed. The sample books of textile and dye manufacturers tell a different story, with page after page of brilliant colors. "In plain goods may be found as many as one hundred and ninety-three different shades of fashionable colors . . . and there are four to six shades of every color, in many qualities," observed *Demorest's Family Magazine* in November 1890.

In real life, the plaid blouse in a black-and-white engraving might have been woven with pink, blue, yellow, and white on an intense black background. The seemingly subtle swirls on a walking skirt were likely a vivid pink, green, or purple against deepest black—itself a feat of modern chemistry. Green, formerly hard to achieve, was everywhere. Among the designs featured in the April 1891 issue of *Demorest's* was an ensemble combining a skirt of embroidered light-green bengaline, a silk-cotton combination; a bodice of green silk twill adorned with steel beads; and sleeves of dark green velvet.

Play up color contrasts, the magazine urged readers: "Almost every shade of every color is used with black, turquoise blue being especially favored. . . . Gray and bright yellow, old pink and bright red, faded rose with bright rose-color, blue and gold, pink and gold, tan-color and stem green, are favorite combinations, and brown is combined with old rose or fern-green, bright French blue or gold."[46] From the most luxurious silk velvets to the humblest cottons, late-nineteenth-century fabrics displayed an unprecedented profusion of colors.

These textiles embody one of the most important developments not simply in the history of fashion but in the history of technology: the synthetic dyes that gave rise to the modern chemical industry. Beginning in the 1850s, the pursuit of new textile colors employed generations of chemists. The demand for dyes offered a career path, challenging problems, and potential riches to some of the era's most inventive minds—much the way information technology attracts people today. Innovations born of dye chemistry altered the balance of political, economic, and military power, produced the first wonder drugs, and gave us plastics and synthetic fibers.

"In the late 19th century," writes a historian of science, "the synthesis of colors brought together scientific knowledge and industrial technology, the research laboratory and the modern business firm. The makers of dyestuffs diversified into photographic supplies, insecticides, rayon, synthetic rubber, resins, fixed nitrogen, and, not least important, pharmaceuticals."[47] Dyes made the modern world.

It all started with industrial waste.

In the nineteenth century, coal gas lit city homes, businesses, and streets. Coke, made by purifying coal in beehive-shaped ovens, powered iron and steel furnaces. Converting coal into these concentrated fuels left behind a sticky, viscous residue called *coal tar*. A sludge of assorted hydrocarbons, this otherwise useless by-product drew the attention of August Wilhelm

Hofmann, a German graduate student who was investigating organic compounds containing nitrogen.

The compounds found in plants and animals—and, it turns out, in coal tar—perplexed nineteenth-century chemists. It wasn't enough to figure out what elements they contained, because the list was the same: carbon, hydrogen, oxygen, and sometimes nitrogen, sulfur, or phosphorous. What made one compound of the same elements different from another? Why did some atoms seem easily replaced while others of the same substance remained fixed? Not until the late 1850s, when August Kekulé began publishing theories on how carbon atoms could form chains or rings, did chemists begin to understand molecular structures. Until then, simply identifying unique compounds was a major challenge.

In his first scientific paper, published in 1843, Hofmann demonstrated that an alkaloid derived from coal tar was identical to three other previously discovered chemicals: one made from benzene, another coal tar product, and two distilled from indigo plants. The four supposedly different substances were actually a single compound. It contained six atoms of carbon, seven atoms of hydrogen, and one atom of nitrogen or, to put it in another way, it included an amino group (two hydrogens and a nitrogen) and a distinctive combination of six carbon atoms and five hydrogens. Hofmann called the compound *aniline*, from the Arabic for "indigo."

Hofmann's discovery held practical promise. It proved that the same chemical found in plants could be made from industrial hydrocarbons. Doctors relied on plant alkaloids for such critical drugs as morphine and quinine, and Hofmann's results fueled hopes that, with enough experimentation, chemists might learn to synthesize these vital substances. Hofmann called aniline his "first love" and devoted much of his life to understanding how it related to other compounds.[48]

In 1845, the young chemist accepted a job as the first director of the new Royal College of Chemistry in London, a school that sought to train professional chemists rather than teach a dab of chemistry to future doctors, lawyers, and engineers. It was a heady time for organic chemistry, with discoveries coming rapidly yet much still unknown. That chemical instruction was hard to come by made it all the more desirable to ambitious young minds.

In his new post, Hofmann taught eager students the experimental techniques pioneered in Germany. Still in his twenties, he immediately became a beloved mentor, with "complete sway over his pupils," as one later recalled.

> It was Hofmann's rule . . . to visit each individual student twice during the day's work, and to devote himself as patiently to the drudgery of instructing the beginner, or of helping on the dull scholar, as he did, delightedly, to the guidance of the advanced student, whom he would skillfully delude into the belief that the logical succession of steps, in making the first investigation which the master had selected for pursuit by the pupil, was the result of skill in research which he had already attained, instead of being simply or mainly the skillful promptings of the great master of original research.[49]

William Perkin (left), who invented the first synthetic dye when he was just a teenager, and his teacher, August Wilhelm Hofmann, who made the critical discovery that aniline, a compound found in indigo plants, was also found in coal tar. (Wellcome Collection)

Capturing Hofmann's spirit, his most famous London pupil remembered how one day when the chemist was making his rounds he picked up the product of a student's successful experiment, put a bit of it in a watch glass he carried with him, and added a caustic alkali. The chemical immediately transformed into a "beautiful scarlet salt." Looking up enthusiastically at the students gathered around him, Hofmann exclaimed, "Gentlemen, new bodies are *floating* in the air."

Entranced by the beauty of chemistry, Hofmann personally preferred pure science. Still, he and the college's supporters hoped the school's research would lead to practical breakthroughs. The early results were disappointing. "None of these compounds have as yet found their way into any of the appliances of life. We have not been able to use them for dyeing calico nor for curing disease," Hofmann admitted to backers in 1849. Within a few years, that would change radically, thanks to a teenager's experiments.[50]

William Perkin entered the college in 1853, when he was just fifteen, and was soon one of the chemical whizzes who sparked Hofmann's delight. Although Perkin's first research project on a coal tar derivative failed, his experimental technique impressed the master, who named him a research assistant. So enthusiastic was Perkin about chemistry that he constructed a small lab in his home, where he could work when school was out. Over Easter vacation in 1856 he made a world-changing discovery.

Like many organic chemists, Perkin wanted to synthesize the antimalaria drug quinine, which came from the bark of a tropical tree. Chemists knew its composition, but they couldn't make it. "Very little was known of the internal structure of compounds," Perkin later explained, "and the conceptions as to the method by which one compound might be formed from another were necessarily very crude."

His first attempt to produce quinine failed. Instead of the colorless compound he hoped for, he got only "a dirty reddish-brown precipitate." Out of intellectual curiosity, he decided to repeat the experiment, starting this time with a compound of Hofmann's beloved aniline. Again, no quinine, just a black precipitate. Curious about what it might be, Perkin tried dissolving the new substance in denatured alcohol. The solution turned a striking purple. Suddenly, the experiment was once again practical. If not a medicine, perhaps the chemical could be a dye.

In a different time or place, an ambitious young chemist might have discarded the failed experiment or investigated the precipitate's composition merely for its own sake. Dye wouldn't have come to mind. But in nineteenth-century Britain, textiles were the most prominent industry and dyes were big business. A colorful solution naturally conjured visions of dye profits—all the more so if, as in this case, the particular shade was having a fashion moment. Perkin tested the mystery solution on cloth. "On experimenting with the colouring matter thus obtained," he later wrote, "I found it to be a very stable compound dyeing silk a beautiful purple that resisted the light for a long time."

Although he knew how to synthesize it, Perkin didn't actually understand the aniline purple he'd created. He didn't yet know its molecular formula, let alone its structure. But he quickly grasped its possible use. "He made the initial breakthrough by using, rather than by attempting to interpret, experimental results," observes a historian of science. "It was, in fact, the only way in which training in organic chemistry could be of value outside the laboratory until the valency and structural theories developed between 1858 and 1865 were harnessed."

After further tests, Perkin contacted a Scottish dyeing firm to probe commercial interest. "If your discovery does not make the goods too expensive," the owner's son replied,

> it is decidedly one of the most valuable that has come out for a very long time. This colour is one which has been very much wanted in all classes of goods, and could not be obtained fast on silks, and only at great expense on cotton yarns. I enclose you pattern of the *best* lilac we have on cotton—it is dyed only by one house in the United Kingdom, but even this is not quite fast, and does not stand the tests that yours does, and fades by exposure to air. On silk the colour has always been fugitive.

That fall Perkin left the college to turn the discovery of what he dubbed Tyrian purple into a commercially viable product.

Like many entrepreneurs, Perkin benefited from his ignorance. If he'd known how hard the venture would prove, he might have avoided it—as, indeed, Hofmann cautioned him to. "At this time," Perkin admitted, "neither I nor my friends had seen the inside of a chemical works, and whatever knowledge I had was obtained from books." Scaling up to industrial production was much harder than producing small quantities of dye on a laboratory bench.

Synthesizing large quantities of the dye, its component aniline, and the chemicals needed to produce aniline from benzene required inventing new industrial equipment. "The kind of apparatus required and the character of the operations to be performed were so entirely different from any in use that there was but little to copy from," recalled Perkin.

While silk drank up the color, cotton resisted it—and the big money was in cotton, especially cotton printing. Several years passed before printers developed reliable ways to fix the dye on cotton without interfering with other colors. Perkin himself devoted much of his time to on-site visits to customers, developing and teaching new techniques for preparing fabric for his product.

The effort paid off. By 1859, the dye, popularly known by the French name *mauve*, was a smashing success—so much so that the satirical magazine *Punch* reported a plague of "mauve measles." Other chemists raced to emulate Perkin, either by copying him directly—he held only a British patent—or inventing their own dyes. With the success of Perkin's invention, writes Simon Garfield in his history of mauve, "the full force of chemical ambition was unleashed."[51]

Within a few years, mauve was out. Another aniline dye, called *fuschine* by its French creator and *magenta* by the English, was in. Pure scientist though he was, even Hofmann eventually got into the dye game, patenting a range of aniline shades that came to be known as Hofmann's violets. As the quest for chemical colors intensified, the new chemical industry grew, particularly in Germany. The dye-driven demand for intermediate chemicals such as aniline and benzene gave rise to new manufacturing plants and, once supplies were easily available, to additional uses for those chemicals, including for pure research. As the dye industry "has utilized the discoveries of chemists," Perkin said in 1893, "it has handed back to them in return new products which they could not have obtained without its aid, and these have served as materials for still more advanced work."[52]

Dye research itself continued, taking advantage of the structural models that began with Kekulé. Chemists learned to synthesize and alter molecules once found only in nature. Molecular clones birthed in German corporate labs shoved aside madder in the 1870s and supplanted indigo by the end of the century. Vast tracts long devoted to these staples were suddenly obsolete. The madder fields of France returned to vineyards.

In India, the transition was particularly abrupt. In its peak year, ending in March 1895, British India exported more than nine thousand tons of indigo dye. A decade later, that volume had plummeted by 74 percent while the revenue received dropped by 85 percent. The reason was the introduction of synthetic indigo in 1897. "These figures present a melancholy record of the decline of an old and important industry," declared a government report. "The unremunerative level to which prices have been forced down by the competition of synthetic indigo has reduced the indigo plantations of Bengal to less than half the area they occupied ten years ago, and over the whole of India the reduction in that period is 66 per cent." By 1914, that number was 90 percent. Chemistry displaced colonies as a source of geopolitical power. Germany was on the rise, and the world would never be the same.[53]

𝕫𝕫𝕫𝕫

Khalid Usman Khatri squats as he dips the length of cloth into one of the seven plastic water tubs arrayed around him on the ground, then kneads the wet fabric on the flat surface of a couple of cinder blocks. Now the fun begins. Grabbing one end of the cloth, Khatri raises the material above his shoulder and repeatedly brings it down onto the hard surface. *Thwack. Thwack. Thwack. Thwack.* Whipping the cloth against the concrete, he beats out the excess dye from one of the three rounds of block printing that will produce an intricate black-and-white design.

A master of the Indian art known as *ajarkh*, Khatri employs traditional techniques in novel ways, designing original printing blocks that give inherited motifs a contemporary edge. Normally, he runs a workshop and doesn't do his own washing. This week, however, he's at the Somaiya Kala Vidya design school, introducing block printing to a couple of foreign amateurs. Thanks to my early-morning food poisoning, he had a few free hours to make something new. So he's been experimenting with an iron-based monochrome instead of the colors that shook the world. He's using a lot of water.

Washbasins, I learn in my week of Indian dyeing, are as essential to the process as dye stuffs, mordants, and carved blocks. Rinse and dump, rinse and dump—tub after tub of water gets hurled into the yard. To my drought-trained Angeleno eyes, it seems like a disturbingly thirsty process. We are, after all, in Adipur, in the desert region of Kutch in central India's far west. Water is actually more abundant in thirsty Southern California. Although Khatri uses the natural dyes beloved by those who want to feel close to the earth, his process doesn't exactly prize resource conservation.[54]

In our eco-conscious age, many people assume that preindustrial life was environmentally benign. But, as we've seen, dyeing has always been a mess: dependent on large supplies of water, fuel, and stinky ingredients. (Indigo that smelled like urine! Bran water that smelled like vomit! Snail carcasses that smelled like rotting flesh!) For millennia, the primary strategy for avoiding the negative side effects has been to make sure dyeing is done *somewhere else*—the other side of town or the other side of the world. People crave the beautiful results but don't want to live next to dye houses.

So I was surprised to discover a major dyeing and finishing plant in Los Angeles, where "not in my backyard" is practically the official motto. Here, water is scarce, air emissions are highly regulated, electricity and labor are expensive—and that's before you get to the taxes.

Nonetheless, "we've figured it out and we're doing really well," says Keith Dartley, one of the owners of Swisstex California. (The other three owners are Swiss, the source of the name.) Founded in 1996, the company started out serving contractors who made private-label apparel in LA and Mexico. Retailers had finally begun imposing quality standards, and, instead of buying whatever was cheapest, contractors needed fabrics that were reliably colorfast and didn't shrink or twist. As established suppliers struggled to keep up, Swisstex built a state-of-the-art facility to meet the new expectations.

Today its original market has largely vanished, decamping to Asia. What's still around is athletic wear—and it's booming. Swisstex dyes, finishes, and, in some cases, manufactures knits for athletic brands, including Nike, Adidas, and Under Armour, and for companies that make plain T-shirts, hoodies, and other staples for customized printing. In 2019, Swisstex expanded capacity by 40 percent at both its original site in LA and a sister operation in El Salvador. The LA plant now dyes and finishes about 140,000 pounds of fabric a day, or roughly three hundred thousand yards. In El Salvador, the capacity is about two-thirds as much. That's a lot of T-shirts.[55]

It could also be a lot of air pollution, water and power consumption, and chemical runoff—the notorious products of industrial dye houses. Asian rivers flowing with the colors of the next fashion season are a journalistic touchstone, and in 2017, the *Hindustan Times* reported that stray dogs in a Mumbai suburb were turning blue from swimming in a local river; the exposé led regulators to shut down a dye house.[56] At a small plant in Surat, an Indian textile center in the western state of Gujarat, my host shows me the up-to-date air pollution equipment that captures fine particles from the coal-fired boiler—and leaves them in a pile on the ground. That might satisfy local regulations, but it would never fly back home.

Complying with California's strict standards, Swisstex burns natural gas, not coal, using special equipment that minimizes emissions. On the back end, it sends exhaust from drying fabrics into a machine known as a thermal oxidizer. Heating the air to 1,200 degrees Fahrenheit, it breaks down any hydrocarbons that may have leached from the fabric, producing carbon dioxide and steam. That fulfills air pollution requirements but isn't the end of the story. The particles actually help fuel the machine, reducing natural gas use. The system also captures the steam to preheat dye water. "Instead of taking room temperature water and heating it for dyeing, it's already hot," says Dartley. "We save a lot of energy there." Per pound of fabric, the company says, Swisstex consumes half the energy of the typical US dye house and much less than most abroad.

Swisstex survives—indeed, prospers—because its efficiency-obsessed owners continually drive down the amount of water, electricity, gas, and labor needed to dye each pound of cloth. Skylights reduce lighting costs and,

Dye containers in Swisstex's Los Angeles test lab, where robots measure out precise amounts to create small quantities of new color formulas, reducing waste and ensuring exact replication. (Author's photo)

because they open, vent hot air. Solutions of salt or soda ash are premixed and ready to go when needed, cutting downtime. Computer-controlled robots precisely align the seams joining fabric bolts in the same dye lot, minimizing distortion and waste. Other machine modifications and process tweaks are invisible to the casual visitor. "It's been twenty-five years of pressing the envelope," says Dartley. "There's no equipment that's in the state we received it." The incremental improvements add up.

Take water consumption. A decade ago, Swisstex used about 5 gallons to dye each pound of fabric. That's an impressively small figure, less than the amount in just one of the washtubs back in Adipur and an unusually low ratio for an industrial plant. A well-run dye facility might easily use 25 gallons; a wasteful one, as much as 75. Even more impressive is that, over the past ten years, Swisstex has cut water usage by 40 percent—from 5 gallons a pound to 3. "We use less water per pound than any dye house on earth," Dartley boasts. That achievement didn't come with a single breakthrough or a new piece of equipment but from hundreds of small improvements across the entire process: better machines, better dyes, more accurate controls.

"You will see sometimes one of my partners out here with a stopwatch, timing things literally in the seconds, to see where can we shave off a little more time," says Dartley. Back in the early 1990s, when the founders worked together at another dye house, it took twelve hours of dyeing to produce dark colors, compared to between four and five today. Saving time means reducing power, which means saving money and, for those who care, carbon emissions.

And, lately, more people do care. "This is something that's become very important *this year*," says Dartley on my visit in September 2019. "This year is the first time where I've begun to see brands and retailers making sourcing decisions on sustainability. Why? Because the consumer is no longer accepting irresponsible environmental practices, and there's so much transparency with the internet." In an intensely competitive industry, environmental credentials now matter. Customers still want their clothes to be attractive, comfortable, and reasonably priced. But eco-friendliness has become fashionable.

Creating colorful textiles with minimal side effects is increasingly possible. But it requires precise controls, advanced technology, and constant improvement. You don't get it by thinking like a nature child. You get it by thinking like a Swiss engineer. Environmentally benign dye technology isn't a lost art. It's something we're still inventing.[57]

Chapter Five

TRADERS

O Wool, noble lady, you are the goddess of the
merchants. All of them bow down to serve you. In your
fortune and your riches, you cause some to mount on
high, and others you cast down to the depths.

—John Gower, *Mirour de l'omme*, c. 1376–1379

LAMASSĪ WAS DOING HER BEST to keep up with the demand for her fine
woolen cloth, fickle though the requirements seemed to be. First her husband asked for less wool in the fabric, and then he asked for more. Why
couldn't he make up his mind? Maybe it was his customers in that distant
country. Maybe they didn't know what they wanted. At any rate, her latest
batch of cloth, or most of it, would soon be on its way. She wanted Pūsu-kēn
to know it was coming. She wanted him to know she was doing her job. She
wanted a little appreciation.

Lamassī rolled a small ball of damp clay between her hands, then flat-
tened and smoothed it into a neat, pillow-shaped tablet, which she cupped in
her left palm. She picked up her stylus and began to write, pressing wedge-
like characters into the wet clay.

Say to Pūsu-kēn, thus says Lamassī

Kulumāya is bringing you nine textiles. Iddin-Sîn is bringing you three textiles. Ela refused to take any textiles and Iddin-Sîn refused to take another five textiles.

Why do you always write to me, "The textiles that you send me each time aren't good!" Who is this person living in your house and denigrating the cloth that I send to you? For my part, I do my best to make and send you textiles so that for every trip at least ten shekels of silver can reach your house.

Her message completed, Lamassī dried the tablet in the sun. She then wrapped it in a gauzy fabric, which she coated with a thin layer of clay. She ran a cylindrical seal along the clay envelope to mark the letter as hers. A messenger would take it to her husband, 750 miles away in the Anatolian city of Kanesh.

Four thousand years old, Lamassī's letter is one of some twenty-three thousand cuneiform tablets excavated from the site in Turkey that was once the city of Kanesh. Almost all found in the homes of expatriate merchants like Pūsu-kēn, these letters and legal documents preserve the practices and personalities of a thriving commercial culture. They are our oldest records of long-distance trade.[1]

From Bronze Age caravans to today's container ships, textiles have always been central to commerce. Clothing the body and furnishing the home, they are at once necessities, objects of beauty, and prized status goods. Fabric is easily transported, fibers and dyestuffs flourish in specific regions, and particular communities develop skills that make their textile products especially desirable. These characteristics all encourage local specialization and its complement, exchange.

In addition, each stage of textile production, from fiber to finished cloth, is usually separated in time and space from the next. Each incurs expenses that have to be covered long before the ultimate sale is realized. Each poses new dangers that accident, natural catastrophe, theft, or fraud will wipe out the goods' value. How do you cope with natural threats—weather, pests, disease—and human malfeasance? How do you know exactly what you're buying? Assuming all goes well, how do you get paid? Commercial civilization depends on answering these questions.

Like spindle whorls and mounds of murex shells, the tablets known as the "Old Assyrian private archives" testify to the central role of textiles in the early history of innovation. Here, the inventions aren't material artifacts or physical processes but "social technologies": the records, agreements, laws, practices, and standards that foster trust, ameliorate risks, and allow transactions across time and distance, even among strangers.[2]

By enabling peaceful exchange, these economic and legal institutions permit larger markets and, with them, the division of labor that leads to variety and abundance. They are as essential to prosperity and progress as anything devised in a workshop or laboratory. Along with the economic benefits come less-material gains, giving humans new ways to think, act, and communicate. And once again, driving the invention we find the desire for textiles.

卪卪卪卪

Lamassī lived in Aššur, on the Tigris River near Mosul in modern-day Iraq. Centuries later, the town would give its name to the Assyrian empire, but in her day it was a modest city-state run by merchants.

Except for donkey harnesses and its women's cloth, Aššur itself produced little. It was instead a commercial hub. From faraway mines to the east came tin, essential for the copper alloy from which Bronze Age tools and weapons were forged. From the south came Akkadians bearing wool cloth, made in workshops by women prisoners and slaves. Raw wool arrived on the hoof, as nomads drove their flocks to the city to be plucked. Aššur's women bought the fleece to spin and weave into their much-desired fabrics, each piece a standard eight cubits wide and nine cubits long, about four yards by four and a half. "A single textile of fine quality," observes Assyriologist Mogens Trolle Larsen, "could easily cost the same as a slave or a donkey."

Aššur was a city of middlemen—the earliest we have records of, although not likely the first. Its merchants purchased tin and textiles and exported them, along with its women's weaving, to Kanesh. Twice a year, avoiding the winter storms that closed the mountain passes, caravans of donkeys made the six-week journey. A single caravan might include wares from eight different merchants, with thirty-five donkeys carrying more than a hundred pieces of cloth and two tons of tin. Some of the wares went for taxes in the two cities and in kingdoms along the way that guaranteed safe passage. The rest were traded for silver and gold. In other correspondence, Pūsu-kēn gives Lamassī an accounting of her textile proceeds: how many pieces went for taxes, how many were sold, what profit he's returning to her, and what money he still anticipates. We have his letter because he kept a copy.

A cuneiform letter from Kanesh discussing the textile trade, circa twentieth to nineteenth century BCE (Metropolitan Museum of Art)

By the time Lamassī picked up her stylus, cuneiform script was a thousand years old. For most of that time, however, writing had been the monopoly of a small class of specially trained scribes, probably a mere 1 percent of the population. Throughout most of human history, literacy belonged to the few, mostly men working for state or religious institutions.

Not so in Aššur.

"In this society of traveling merchants," writes Larsen, "it was essential that men and women taking part in the commercial activities had a certain mastery of the script. They had to be able to read a letter when they were far away in a village where there was no professional scribe, or they would find themselves in a situation where a letter contained confidential information which should not be broadcast or even seen by outsiders." For the Old Assyrians, letters were a critical technology.

Assyrian merchants needed to send instructions between Aššur and Kanesh and between Kanesh and the surrounding towns, where their agents sold textiles and tin. They needed to record orders, sales, loans, and other contracts. They needed the flexibility and control that come with literacy.

Over time, these pragmatic merchants simplified the cuneiform script, making it easier to learn and write. They invented a new form of punctuation that helped them skim documents quickly. Some wrote well, others poorly. But in this society of long-distance traders, most men and many women were literate.[3]

Trade requires clear communication, particularly if the business owner doesn't conduct every negotiation personally. Consider Pūsu-kēn. He first went to Kanesh as the agent of an older merchant in Aššur and, even as his own ventures grew, he continued to work on behalf of various traders back home. When their textiles and tin arrived in Kanesh, he needed to know what to do with the goods.

One option was to sell the shipment immediately in the city's marketplace. "Have them sell my goods for cash on delivery at what price they can get," a cash-hungry merchant wrote to Pūsu-kēn. "Give instructions that they must not release the goods on credit to an agent!" In this case, the silver needed to return right away, even if a quick sale meant accepting a lower price.

The alternative was for Pūsu-kēn to sell the textiles and tin to an agent who agreed to pay after a certain period of time. The debt contract would be sealed in an envelope on which its text was repeated; the envelope was broken when the debt was repaid. "Take the tin and the textiles en bloc," another merchant in Aššur instructed Pūsu-kēn,

> and sell the goods [on credit] either on short terms or on long ones, as long as a profit can be assured. Sell it as best you can and then inform me in a letter about the price in silver and the terms.

An agent buying on credit typically paid about 50 percent more than could be gotten in Kanesh's marketplace. He then peddled the goods in out-lying towns, where prices were higher. By providing working capital, the arrangement gave him time to turn a profit himself despite paying a premium for the goods—a win for both sides.

Assuming, of course, that the agent paid his debt. He might abscond with the goods, never returning to Kanesh. He might fail to turn a profit and simply refuse to pay. He might be robbed or injured or even die. Selling on credit entailed risks, and letters from Aššur often urged the recipient to find an agent "as reliable as you yourself." With the written contract, a merchant could take the debtor to court, assuming he could find him. But then, as now, dealing with someone who upheld his end of the bargain was far preferable.[4]

Letters are such an old technology that we take them for granted. But they were crucial to long-distance trade. Articulating, transmitting, and preserving the sender's instructions, letters were "the tools that allowed a merchant to project his authority over his goods and money across space," writes a historian. She's referring to the Jewish merchants who traded textiles, dyestuffs, and other goods across the Islamic Mediterranean in the eleventh century CE.[5] But the description could apply to any era before the telephone. When commerce stretched over time and distance, written correspondence—and the literacy it required—came with it.

▨▨▨▨

The people of Turfan, an oasis city in present-day Xinjiang in northwest China, dressed their dead in garments, shoes, belts, and hats made not of cloth or leather but of discarded contracts and written documents. Today, these recycled papers constitute a remarkable, if somewhat random, record of the institutions and customs of the city's polyglot inhabitants. They include the oldest surviving Chinese-language contract—the purchase of a coffin in exchange for twenty bolts of degummed silk (*lian*) in 273 CE. In another contract, from 477, a Sogdian merchant buys an Iranian slave for 137 bolts of cotton, the earliest written record of cotton in the region. These weren't simply barter transactions. In Turfan, cloth was a critical social technology: money, denominated in standard bolts, just as silver was money in Aššur.[6]

When China conquered Turfan in 640, the new rulers further entrenched cloth as currency, using it to pay soldiers and buy provisions. A ledger left by a Chinese soldier named Zuo Chongxi, who was also a wealthy farmer, records how many bolts of silk he used for purchases including horses, a sheep, rugs, and horse feed. He reserved coins for smaller transactions. For a fifteen-year-old slave, he paid six bolts plus five coins. Silk bolts were the big bills, coins the change.[7]

Suffering from a chronic shortage of coins, especially in rural areas, the Tang dynasty (618–907 CE) encouraged textiles as an alternative. In 732, the government declared bolts of hemp and silk to be legal tender, meaning they must be accepted as payment. In 811 it ordered citizens to make all large purchases with either textiles or grain rather than coins. Most important, it collected taxes in standard measures of grain and standard bolts of silk or hemp. Its armies ate the grain, but the textiles circulated as money. Paid in bolts of silk and hemp, soldiers and officials spent their salaries in local markets; shopkeepers then made their own purchases with the cloth money. Coins served as accounting units, but bolts of cloth were the everyday medium of exchange.

A story recounted by the ninth-century writer Li Zhao captures the situation. On a winter day, a cart filled with heavy pottery got stuck in the snow and ice, blocking a narrow road. Hours passed, and the crowd of frustrated travelers behind the cart grew increasingly large. Darkness threatened.

> Then, one traveller, a certain Liu Po, came forward, his whip in his hand and asked: "How much are the jars on the cart worth?" The answer was "7,000 to 8,000 coins," at which Liu Po opened his bags, took out some degummed *lian*-silk, paid the price, ordered his servant to climb on to the cart, loosen the ropes, and push the jars over the cliff. Moments later, the lightened cart could move forward. The crowd cheered and went on their way.

As a historian observes, the story demonstrates that traveling merchants routinely carried silk bolts to use as money and that they could quickly calculate the value of coins in silk: "The speed at which the transaction was completed shows that converting between silk and coins was a generally accepted practice at the time and a skill which much of the populace possessed."[8]

In preindustrial economies, textiles have many of the characteristics essential to a good currency. They're durable, portable, and divisible. Bolts can be produced in standard sizes and uniform quality. The quantity is limited, because cloth takes a long time to produce and it flows out of the money supply as it's transferred to everyday use, thereby avoiding inflation.

Although we tend to think of money as something established by central authorities, as silk money was in Tang China, that need not be the case.

Elsewhere in the world, textile currency emerged out of commercial usage, supported, but not created, by law.

The Icelandic tale of Audun begins in the early summer in the mid-eleventh century when a Norwegian merchant named Thorir arrives on the island's northwestern Westfjords peninsula.[9] Living in a land inhospitable to forests or farming, Icelanders relied on imports for timber and grain. They paid for these goods in the same currency they used locally: a woolen twill cloth called *vaðmál* (or *wadmal*), pronounced *wahth-mall*. Thorir could sell his goods in Iceland and return with a ship laden with textiles. But there was a problem. The customers didn't have enough cash—*vaðmál*—on hand.

"If the Norwegian was to get paid for his flour and timber, the Icelandic buyer was unlikely to have enough cloth woven until later in the summer at best," explains a legal historian and Icelandic saga scholar. "The merchant would have to wait until you literally *made* your money to pay him and not infrequently the merchant had to stay the long winter to get his payment." Meanwhile, the grain might go bad.

Fortunately for Thorin, the story's Icelandic hero Audun identifies creditworthy customers. If Thorin gives them grain now, he can reliably expect cloth in time to set sail in the late summer. As a reward for his credit-reporting services, Audun gets passage on the ship, setting in motion the tale's events.[10]

Iceland's *vaðmál* wasn't just a commodity. Woven to specific standards, it was a legally recognized medium of exchange and store of value, the primary form of money during Iceland's Commonwealth Period (930–1262 CE). As a unit of account, the third function of money, a piece of *vaðmál* two ells wide and six ells long (about a yard by three yards) was "ubiquitous as a measure and medium of exchange in Icelandic legal texts, sales accounts, church inventories, and farm registers into the seventeenth century," writes anthropological archaeologist Michèle Hayeur Smith.[11]

The archaeological evidence backs up the written records. Microscopically examining more than thirteen hundred archaeological textile fragments, Hayeur Smith found clear indications of cloth becoming money. The material from the Viking Age, before 1050, includes many different weave structures and widely varied thread counts. Medieval fragments, by contrast, are much more uniform—overwhelmingly the dense twills recognized as legal money. The analysis, she writes, reveals "such degrees of standardization and ubiquity that one can only conclude that cloth truly had become a unit of measure, a type of 'legal cloth currency' produced and circulated among households of all ranks across the island." In the Middle Ages, "Icelanders were weaving money in abundance."[12]

In West Africa, too, merchants at least as far back as the eleventh century used textiles to create the currency they needed to conduct trade. For many West African fabrics, narrow strips are sewn together to form a larger textile, which is worn as a single piece. (Kente cloth is one example.) Unlike colorful textiles for apparel, a strip intended as currency would be left undyed and wound into a tight, flat coil as it came off the loom. Merchants could roll such coils on the ground, sling them on either side of a pack animal, or carry them flat on the head with other goods added on top. Because weaving widths varied from place to place, if a market attracted more than one type, traders established a standard rate of exchange. A given strip length, usually that of a woman's wrapper, would be the primary monetary unit, with a full cloth forming a larger denomination.

Although African currency cloth functioned primarily as money, it did have a consumer market among the poor and desert dwellers to the north, who had no cotton. "Cloth currency therefore always had a certain 'one-way' character," writes a historian. In east–west trade, its value stayed essentially the same. Going north, however, a unit of cloth bought more; going south, it bought less. Traders adjusted their travel expenditures accordingly.

> A merchant from Upper Volta, for example, going to Timbuctu to buy salt with cloth produced in his home area, would use cloth to pay his way on the northward journey; but on the return journey he would prefer to use salt which appreciated in value as it moved southward, even if he had first to sell it for local cloth money.

The same was true of silver and gold flowing from the Americas, where it bought less, to Europe and Asia, where it bought more. Cloth money was actually more self-regulating and less prone to shortages or inflation than metallic currencies. When its value rose, weavers would make more. If it became less valuable, consumers would take more. The result was a fairly constant value over time, set by the cloth's price as a commodity.[13]

Money is a self-perpetuating social convention, a token that we trust will be valuable in future exchanges. If buyers and sellers, courts and tax authorities accept textiles as payment, they are money.

☒☒☒☒

In the late thirteenth century, the merchants of Northern Italy began to organize their business in a new way. Instead of making the month-long journey across France to the great international trade fairs of Champagne, they stayed home, arranged for partners or agents to live in the area full-time, and sent goods back and forth with specialized carriers. This division of labor, which Pūsu-kēn might have recognized, was part of what has been called the commercial revolution of the thirteenth century.

At first, business at the fairs increased even as fewer goods changed hands there. "An Italian could agree in Champagne to purchase a certain number of rolls of Flemish cloth of a specified quality, and they could be transported directly from Flanders to Italy, without necessarily passing through the town where the trade took place," explains a historian. Merchants soon realized that they could skip the fairs altogether by opening outposts in the cities where they did most of their business, including Paris, London, and Bruges. By 1292, Paris counted six Italian businessmen among its seven largest taxpayers.[14]

With less face-to-face contact, letters and record-keeping grew increasingly important. Writing to his mother from the Valencia office of his Florentine family's far-flung enterprise, sixteen-year-old Lorenzo Strozzi reported that he copied twelve letters a day. "I write so fast that you would marvel, faster than anyone at home," he wrote in April 1446. By serving as a fifteenth-century copy machine, young Lorenzo learned the family business and the conventions of commercial correspondence. Describing to his mother the fabrics and fashions favored by Catalan ladies, his letters demonstrate a textile merchant's discerning eye. Good letter writing, in both content and manner, was an essential mercantile skill.[15]

With the growth of long-distance operations came another essential social technology: regular mail service. In 1357 Florentine merchants banded together to create the *scarsella dei mercanti fiorentini*, named for the *scarsella*, or leather messenger bag. They hired couriers and horses to make regular trips from Florence and Pisa to Bruges and Barcelona. (The Bruges route also stopped either at Milan and Cologne or at Paris.) Merchants in other cities followed the Florentine example, and by the turn of the century, *scarselle* ran from Lucca, Genoa, Milan, and Lombardy. In time, Barcelona, Augsburg, and Nuremberg emulated the Italian model.[16]

Traveling by *scarsella*, a letter took about a month to go from Bruges or London to the port cities of Italy and Spain. (Ships were faster but made only two trips a year.) Merchants wrote at least that often. "The absence of correspondence for two months was very rare. Generally merchants would complain and demand more letters if the absence of letters continued for over one month," writes historian Jong Kuk Nam. He analyzed commercial correspondence from the enormous archives left by Francesco di Marco Datini, who ran his multinational textile and banking ventures from Prato, near Florence. With letters constantly flowing in from multiple commercial centers, Bruges became not only a hub for wool and linen but also, Nam writes, "the most important center of news and information in Northern Europe."[17]

Letters carrying commercial information traveled especially quickly between Italian cities. On March 7, 1375, the Venetian silk merchant Giovanni Lazzari answered the February 26 letter from Giusfredo Cenami, a fellow merchant in Lucca. Before getting down to business, Lazzari commented on Cenami's letter, leaving future historians a record of the mail schedule. "You said that you received four of my letters in two days," he wrote. "I send them to you in the usual way, on Wednesday and Saturday." Mostly a market report, Lazzari's letter included silk prices, foreign exchange rates, and a fashion update ("At present, young Venetians have begun to dress in the Florentine style").[18]

Thanks to regular couriers, a historian writes, "Florentine, Lucchese, Pisan, Venetian, Genoese and Milanese businessmen abroad were able to bargain in the light of accurate knowledge of their markets and have supplies sent to them to meet known demand." As evidence, the Datini archives preserve nearly fifty years of commodity price lists "from places as widely separated as Damascus and London."[19] Carried by letter, regular business intelligence supported the fortunes, many of them based on textiles, that funded the humanist works and artistic treasures for which we remember the Italian Renaissance.

In 1479, a few months shy of his eleventh birthday, Niccolò Machiavelli left the school where he'd learned to read and write and went to study with a teacher named Piero Maria. The future author of *The Prince* spent the next twenty-two months mastering Hindu-Arabic numerals, arithmetical techniques, and a dizzying assortment of currency and measurement conversions.[20] Mostly, he did word problems like these:

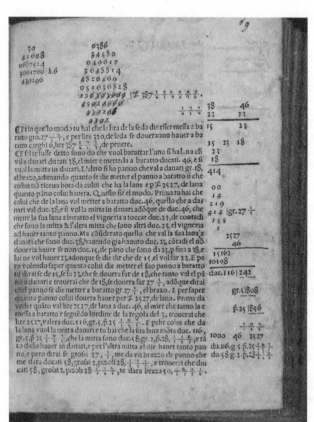

A page from Pietro Borgo's 1561 Libro de abacho, *demonstrating how to solve a word problem about barter versus specie payments for cloth and wool (Turin Astrophysical Observatory via Internet Archive)*

If 8 braccia of cloth are worth 11 florins, what are 97 braccia worth?

20 braccia of cloth are worth 3 lire and 42 pounds of pepper are worth 5 lire. How much pepper is equal to 50 braccia of cloth?

One type of problem reflected the era's shortage of currency. Goods that would sell for one price in coins cost a premium if the buyer paid with other goods. (These problems assume familiarity with trading conventions and therefore present ambiguities to the modern reader.)

Two men want to barter wool for cloth, that is, one has wool and the other has cloth. A canna of cloth is worth 5 lire and in barter it is offered at 6 lire. A hundredweight of wool is worth 32 lire. For what should it be offered in barter?

Two men want to barter wool and cloth. A canna of cloth is worth 6 lire and in barter it is valued at 8 lire. The hundredweight of wool is worth 25 lire and in barter it is offered at such a price that the man with the cloth finds he has earned 10 percent. At what price was the hundredweight of wool offered in barter?

Others were brainteasers dressed up in ostensibly realistic detail.

A merchant was across the sea with his companion and wanted to journey by sea. He came to the port in order to depart and found a ship on which he placed a load of 20 sacks of wool and the other brought a load of 24 sacks. The ship began its voyage and put to sea. The master of the ship then said: "You must pay me the freight charge for this wool." And the merchants said: "We don't have any money, but take a sack of wool from each of us and sell it and pay yourself and give us back the surplus." The master sold the sacks and paid himself and returned to the merchant who had 20 sacks 8 lire and to the merchant who had 24 sacks 6 lire. Tell me how much each sack sold for and how much freightage was charged to each of the two merchants?[21]

Along with their famed humanist arts and letters, the mercantile cities of early modern Italy fostered a new form of education: schools known as *botteghe d'abaco*. The phrase literally means "abacus workshops," but the instruction had nothing to do with counting beads or reckoning boards. To the contrary, a *maestro d'abaco*, also known as an *abacist* or *abbachista*, taught students to calculate with a pen and paper instead of moving counters on a board.

The schools took their misleading name from the *Liber Abbaci*, or *Book of Calculation*, published in 1202 by the great mathematician Leonardo of Pisa, better known as Fibonacci. Brought up in North Africa by his father, who represented Pisan merchants in the customs house at Bugia (now Béjaïa, Algeria), the young Leonardo learned how to calculate using the nine Hindu digits and the Arabic zero. He was hooked.

After honing his mathematical skill as he traveled throughout the Mediterranean, Fibonacci eventually returned to Pisa. There he published the book that enthusiastically introduced the number system we use today. "This method perfected above the rest," he wrote in the introduction, "this science is instructed to the eager, and to the Italian people above all others." Although written in Latin, the language of scholars and churchmen, the book is full of commercial problems.

"It was Leonardo's purpose to replace Roman numerals with the Hindu numerals not only among scientists, but in commerce and among the common people," writes the mathematician who translated *Liber Abbaci* into modern English. "He achieved this goal perhaps more than he ever dreamed. Italian merchants carried the new mathematics and its methods wherever they went in the Mediterranean world."[22] Just as the alphabet once journeyed with Phoenicians carrying Tyrian purple, so calculation traveled with silk and woolen cloth. Once again, the textile trade gave the world new ways to think and communicate.

Fibonacci's novel methods of pen-and-paper reckoning were ideal for businessmen who wrote lots of letters and needed permanent account records. By the late thirteenth century, specialized teachers had begun to teach the new system and to produce handbooks in the vernacular. Consistent sellers, the books served simultaneously as children's textbooks, merchants' reference tools, and, with their brain-teasing puzzles, recreational materials.

The hundreds of manuals the abacists published include the earliest known printed book of mathematics, the *Treviso Arithmetic* (originally *L'Arte del'Abacho*) of 1478, as well as a work by the painter Piero della Francesca. (His well-known book on perspective incorporated the new math.) The most encyclopedic manual, Luca Pacioli's *Summa de Arithmetica Geometria Proportioni et Proportionalità*, published in 1494, was the first work to popularize a complementary social technology: double-entry bookkeeping.[23]

Appealing to owners of far-flung enterprises, the new accounting improved security against embezzlement while providing better information on the state of the business. "It demanded greater care and accuracy by the clerks," write two business historians,

> providing arithmetic checks from periodic balancing, and it permitted the division of labor among several clerks of varying qualification. It provided balance-sheet data, separated capital from revenue accounting, and

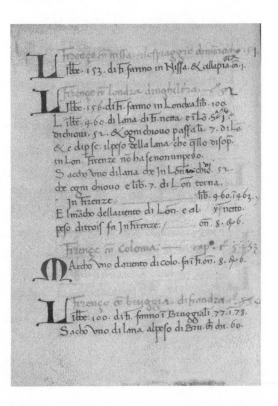

A page from Giorgio di Lorenzo Chiarini's 1481 Libro che tracta di marcantie et usanze di paesi *(Book Concerning the Trade and Customs of Various Places), a guide for converting currencies, weights, and measures across countries. Its entire text was included in Luca Pacioli's* Summa de Arithmetica. *(Temple University Libraries, Special Collections Research Center, via Philadelphia Area Consortium of Special Collections Libraries and Internet Archive)*

introduced useful concepts such as accruals and depreciation. Above all, it gave the owners of the enterprise a much-improved system of control.[24]

Double-entry bookkeeping used pen-and-paper reckoning with Hindu-Arabic numerals, boosting demand for merchants and clerks trained in the new math. Artisans, too, recognized its usefulness for solving everyday problems.

Hence the spread of *botteghe d'abaco*, beginning in Florence in the early fourteenth century. They "mark the first appearance in the West of schools devoted exclusively to the study of mathematics," writes historian of mathematics Warren Van Egmond, "and they were surely the first to teach mathematics on an elementary and practical level."

From the abacists' classrooms, future merchants and artisans typically graduated to apprenticeships and work. But a grounding in commercial math was also common for those like Machiavelli, who were destined for higher education and a career of statesmanship and letters. In a society based on trade, cultural literacy included calculation.

As they drilled generations of children on how to convert hundredweights of wool into braccia of cloth or to allocate the profits from a business venture to its unequal investors, the abacists invented the multiplication and division techniques we still use today. They made small but important advances in algebra, a subject universities scorned as too mercantile, and devised solutions to common practical problems. On the side, they did consulting, mostly for construction projects. They were the first Europeans to make a living entirely from math.

In his seminal 1976 study of nearly two hundred abacus manuscripts and books, Van Egmond emphasizes their practicality—a significant departure from the classical view of mathematics, inherited from the Greeks, as the study of abstract logic and ideal forms. The abacus books treat math as *useful*. "When they study arithmetic," he writes, "it is to learn how to figure prices, compute interest, and calculate profits; when they study geometry it is to learn how to measure buildings and calculate areas and distances; when they study astronomy it is to learn how to make a calendar or determine holidays." Most of the price problems, he observes, concern textiles.[25]

Compared to scholastic geometry, the abacus manuscripts, with their problems about trading cloth for pepper, are indeed down to earth. But they don't scorn abstraction. Rather, by applying the science of patterns to the everyday concerns of business, they wed abstract expression to the physical world. The transition from physical counters to pen-and-ink numerals is in fact a movement *toward* abstraction. Symbols on a page represent bags of silver or bolts of cloth and the relationships between them. Students learn to ask the questions, How do I express this practical problem in numbers and unknowns? How do I better identify the world's patterns—the flow of money in and out of a business, the relative values of cloth, fiber, and dyes, the advantages and disadvantages of barter over cash—by turning them into math? Mathematics, the abacists taught their pupils, can model the real world. It does not exist in a separate realm. It is useful knowledge.

ᛙᛙᛙᛙ

Thomas Salmon had a problem. As a tax collector in Somerset, Salmon had amassed thousands of pounds of gold and silver that needed to get to London. But the England of 1657 had no checking accounts, wire transfers, or armored cars. Physically traveling with that much specie was difficult and dangerous. What was Salmon to do?

He took the coins to local cloth makers, known as *clothiers*. In return, they gave him slips of paper called *bills of exchange*. These bills worked like checks, but instead of drawing on a bank, they told a London businessman named Richard Burt to give Salmon cash. Burt was a *factor*, or middleman; he bought woolen cloth from scattered producers and sold it to London merchants, taking a commission from the sale.[26]

When he sold their goods, Burt kept the clothiers' credits on his books, and they drew down their accounts with bills of exchange. A Somerset clothier could buy household provisions from a local merchant and pay with a *bill of exchange*. The merchant would cash the bill on a trip to London or, more likely, use it to pay his own suppliers who had dealings there. Accepting coins from the tax man was yet another way for clothiers to cash in their credits. Salmon would carry the bills of exchange to London, exchange them for specie at Burt's, and deposit the money at the treasury. An

Bill of exchange, issued by Diamante and Altobianco degli Alberti to Marco Datini and Luca del Sera on September 2, 1398 (akg-images / Rabatti & Domingo)

institution created to serve the textile industry had become crucial to the finances of the British Crown.[27]

Originating with Italian textile merchants in the thirteenth century, bills of exchange have been called "the most important financial innovation of the High Middle Ages."[28] They started as a way for merchants to transfer proceeds from the fairs at Champagne (and later from other markets) back to the home office. Written in a kind of shorthand, these slips of paper were essentially form letters telling an agent, usually a bank, in another city to pay someone a certain amount; when a merchant issued a bill of exchange, his local bank sent a notice to its foreign branch, telling it to honor the bill when presented. Bills of exchange were not official, state-sanctioned documents, designed in advance but, rather, social technologies that evolved through trial and error. Their usefulness depended on connections and trust.

As merchants built up networks of offices in multiple places, bills of exchange became increasingly flexible. By the early fourteenth century, you could cash one in most major cities in western Europe. Whether to buy wool or pay armies, coins no longer needed to be hauled over land and sea. "Bills of exchange," historian Francesca Trivellato writes, "were the invisible currency of early modern Europe's 'international republic of money.'"[29]

Although bills began as a way to easily transport funds and convert foreign money, they quickly evolved other uses as well. For starters, they addressed the shortage of currency by enabling many more transactions with the same amount of specie. In modern economic parlance, they increased the *velocity* of money, not its supply. "The net quantity of silver transported from Bruges to London or Paris to Florence, or of gold from Seville to Genoa did not diminish as a result of the development of bills of exchange," writes a historian, "but the amount of business was increased out of all proportion."[30]

To see why, consider two hypothetical English businessmen. The first (John) exports raw wool, selling it to a Florentine merchant (Giovanni) for a bill of exchange payable in London. The second (Peter) imports silk fabric, buying it (from Piero) with a bill of exchange payable in Florence. On the banks' books, the two bills can be offset against one another, with only the difference actually changing hands as currency. A small supply of coins can thereby enable many more exchanges. "Such a system could be incredibly

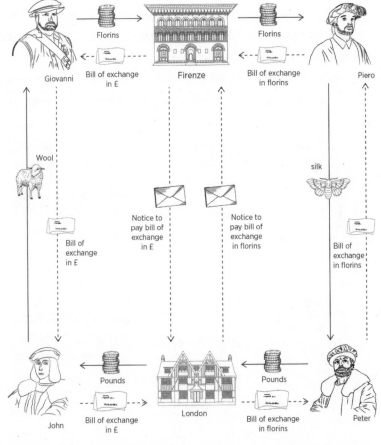

Offsetting bills of exchange allowed many more transactions with limited amounts of specie. The money flows start with John's sale of wool (lower left) and Piero's sale of silk (upper right). (Joanna Andreasson)

efficient," writes economist Meir Kohn. "For example, between 1456 and 1459, one bank in Genoa received 160,000 lire in payments from abroad in bills of exchange, and only 7.5% of this amount was settled in cash: the remaining 92.5% was settled in bank."[31]

Bills of exchange also provided credit. In its simplest form, they gave users a float. A bill was made payable not immediately but after a certain period, or *usance*, from its date of issue. The usance was somewhat longer than the usual travel time between the two cities, ensuring that notice to honor the bill could reach the payer. For a bill written in Florence, a 1442 handbook listed the usance for Naples as twenty days; for Bruges, Barcelona, or Paris, as two months; and for London, as three months. For the same destinations, the estimated courier times were eleven to twelve days, twenty to twenty-five, and twenty-five to thirty days, respectively. The cushion added an extra grace period to the short-term loan.[32]

Over time, merchants figured out ways to turn bills of exchange into overt loans. In a common but oft-condemned practice known as *dry*

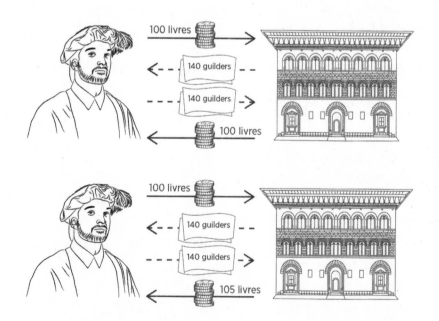

Two forms of dry exchange: In the first, the grace period (usance) on a bill of exchange is used to create a no-interest loan. In the second, exchange rates are changed to charge implicit interest. (Joanna Andreasson)

exchange, the first bill of exchange was paid not in cash (or an account offset) but with a new bill of exchange that simply reversed the original one. This paper swap created an interest-free loan twice as long as the usance. In the fifteenth century, Venetian traders used London as a center for exchanging bills, thereby providing six-month loans. Lenders could extend the terms simply by adding multiple round-trip exchanges.

With a slight variation, dry exchange could dodge bans on charging interest. The trick was to alter the exchange rate on the return bill. If, for instance, a merchant in Bordeaux exchanged 100 livres for an original bill payable for 140 guilders in Amsterdam, the return bill might repay the 140 guilders with 105 livres in Bordeaux. But not all dry exchange involved such ruses. Some transactions were straightforward loans. Indeed, observes Kohn, "as usury restrictions were weakened or eliminated in the sixteenth century, the popularity of the bill of exchange [as a credit instrument] continued undiminished."[33]

Handling bills of exchange drew many a textile entrepreneur officially or unofficially into banking. Francesco Datini, whose archives include more than five thousand bills of exchange, was primarily in the wool trade, but in 1399 he opened a bank in Florence. Its services were state of the art. Along with issuing and accepting bills of exchange, the bank offered "endorsements (*avalli*), guarantees (*fideiussioni*), and correspondence accounts in one or more currencies," writes Datini's biographer. "For payments to third parties, cheques, which were only just coming into use, were freely accepted."

Unlike the city's famous Medici or Alberti banks, Datini's enterprise lent only to private parties, avoiding church and state. Although it prospered, the bank lasted barely three years, succumbing when the Black Death killed the partner who handled the day-to-day business. In tribute to his financial ventures, the nineteenth-century statue of Datini in Prato's main piazza depicts him holding bills of exchange.[34]

The most significant banking operations built on textiles were those of the Fuggers of Augsburg, a wool and linen capital in what is now southern Germany. By his death in 1408, Hans Fugger, who came to the city from a small village in 1367, employed fifty looms. His son Jakob, himself a master weaver, expanded into trading textiles and spices and began issuing bills of

Jakob Fugger (1459–1525), known as Jakob the Rich, built his family's textile business into a banking empire. Woodcut after a silverpoint drawing by Hans Holbein the Elder (iStockphoto)

exchange. He sent his own son Jakob to Venice, where he learned the latest in business practices, including double-entry bookkeeping.

With his brothers, Jakob II built the family's banking operations throughout Europe, often accepting mines and mining rights as security and collecting when Europe's princes failed to pay up. Textile proceeds provided the start-up money to develop highly profitable operations extracting silver, mercury, copper, and tin. The Fuggers parlayed their loans to the Holy Roman Emperor into considerable political influence. They enjoyed a monopoly on transferring money collected from the Catholic Church's sale of indulgences in Germany and Scandinavia back to Rome. Leveraging money and practices from the textile trade, Jakob the Rich, as he was known, amassed the largest fortune of the age.[35]

On a more modest scale, during Britain's commercial expansion in the seventeenth and eighteenth centuries, textile merchants dealing in bills of exchange served the role of country bankers. Take Thomas Marsden, who produced fustian, a popular cloth with a linen warp and cotton weft.

Operating in the late seventeenth century from Bolton, near Manchester, Marsden also maintained an office in London. The business there bought and sold raw materials and cloth, but its primary functions were financial.

By this time, bills of exchange had become negotiable. You could transfer a bill originally made out to you simply by signing the back. The signature conveyed the legal obligation for you to make good on the underlying debt if the bill couldn't be cashed. Once bills of exchange were negotiable, they became more liquid. If you needed cash, you could sell your bills at a discount off their face value, just as bonds change hands today. Or you could issue a new bill of exchange and sell it at a discount to a money broker to redeem later. At least in theory, there was no limit to the number of times a bill could be endorsed, passing from one owner to the next.

"The product of this evolutionary process—the discounting of negotiable bills of exchange—was a financial invention of enormous economic importance," writes Kohn. "Indeed, in the seventeenth and eighteenth centuries it was to become the foundation of modern commercial banking."[36]

Marsden's office in London maintained "great and considerable sums of money," with which it cashed his bills of exchange and made loans by buying other merchants' bills at a discount. (For a month's loan, Marsden typically charged five shillings per £100, or an annualized interest rate of about 3 percent.) He also served as a "Retorner of the Revenue," conveying tax payments to the capital. Sometimes, he traded specie for discounted bills payable in London, à la Thomas Salmon. At others, he transported coins to the capital hidden in his packs of fustians. People trusted his good name.[37]

"Where buyer and seller were remote from each other in space, or not well acquainted, there were advantages in making use of the services of men like Marsden," writes an economic historian, "for a bill drawn on a London firm of wide repute would pass in almost any part of the country. It is not easy to say at what stage any particular intermediary had ceased to be a merchant and become a banker: suffice it to say that Lancashire had banking facilities long before it had specialist bankers."[38]

Negotiability made bills increasingly useful in everyday commerce, outside of specialized money markets. Although no one had to accept them as payment—bills of exchange were not legal tender—if people trusted the signatories, they were nearly as good as cash. "As material artifacts, they had no intrinsic value," observes Trivellato. "Their monetary worth was the measure of the credibility assigned to the chains of signatories who backed them, rather than any sovereign authority."[39]

Sometimes that trust failed.

In 1788, the largest calico printer in Lancashire, Livesey, Hargreaves and Co., went bankrupt, defaulting on £1.5 million in debt. Its collapse rocked the whole region, with the economic trauma going beyond the immediate effects on employment. The textile printer had routinely paid with its own bills of exchange, which circulated as currency in the local area. Weavers, farmers, shopkeepers—people from every line of work—depended on the now-worthless paper. Many were wiped out. One Manchester bank failed, while another suffered a run. The company's "failure convulsed the whole country for some time," reported a nineteenth-century chronicle.[40]

Yet despite their risks, bills of exchange endured, fading from everyday commerce only when superseded by currencies from central banks. As late as 1826, a Manchester banker testified to their continuing popularity, telling a parliamentary inquiry that he'd seen £10 bills of exchange circulating with a hundred or more signatures. "I have seen slips of paper attached to a bill as long as a sheet of paper could go," he said, "and when that was filled another attached to that."[41]

The committee also heard testimony from a representative of a Scottish bank with the peculiar name of the British Linen Company. Founded in 1747 as a cloth manufacturer, it entered the banking business just a couple of decades later, taking advantage of its many local branches.

Along with the business's insatiable need for working capital, bills of exchange help to explain why so many people who started out as textile merchants wound up as bankers.[42]

⚂⚂⚂⚂

In November 1738, clothier Henry Coulthurst informed weavers that he was cutting their piecework rates and would henceforth pay them in goods rather than cash. Needless to say, they were upset. Food prices were rising, and lower wages meant hunger and want.

Over three days in December, the weavers rioted. They smashed Coul-thurst's mill, wrecked his home, and "drank, carried out, and spilt, all the Beer, Rum, Wine and Brandy in the cellars." They returned the following day to demolish Coulthurst's house, reducing it to rubble, then attacked tenant cottages on the property. On the final day, a Friday, they marched triumphantly through the town of Melksham, in the county of Wiltshire in southwest England. Army troops arrived Sunday night, deterring fur-ther violence. Thirteen men were arrested; one was acquitted and three were eventually hanged.[43]

Like similar disturbances in our own day, these largely forgotten riots, which occurred well before mechanization, prompted public soul-searching and heated debate. Who was at fault for the breakdown in public order? Un-scrupulous clothiers or unreasonable workers? Was the violence justified or, if not, at least understandable?

Times were tough in the woolen business.[44] Customers at home were gravitating toward lighter fabrics and competition abroad was intensifying. Everyone, suggested one observer, had reason to feel aggrieved.

> It is a Grievance to one to want Bread, and hear the piercing Cries of starv-ing Children;—It is a Grievance to another not to be paid his just Wages;—It is a Grievance to another to be forced to give more for the Necessities of Life than their Value, or the Market Price;—and it is a Grievance to an-other to have his House, Buildings, or Necessaries of Trade, destroy'd by a riotous Mob.[45]

Many found it hard to fault either side. Fortunately, there was an alterna-tive: blame the middlemen.

In this case, the supposed villains were the factors who represented cloth-iers in London. "The Sufferings of the Poor employ'd in the working of Spanish Wool, are not owing to the Unmercifulness of Clothiers," asserted a commentator writing under the name Trowbridge, "but the Tyranny of the Blackwell-Hall Factor, who, tho' originally but the Servant of the Maker, is now become his Master, and not only his, but the Wool-Merchant's and Draper's too." Unlike hard-working weavers and clothiers, he charged, factors grew "rich without any Risque and with very little Trouble." They were "useless Drones in the human Hive."[46]

Here we see the dark side of social technologies. Intangible and routine, they lack physical indications of value and are often dismissed as insignificant or condemned as malign.

A clothier was himself a middleman, coordinating cloth production through a putting-out system that provided working capital and marketing. He bought wool, which he turned over to contractors to clean, card, and spin. He then took the yarn to weavers and gave them specifications for what he wanted made. The same went for dyeing and finishing. The clothier bore the cost of materials and paid workers at each stage for their labor.

When the cloth was done, the clothier took it to Blackwell Hall in London, which was the only place in the city where non-Londoners were allowed to sell textiles. The Blackwell Hall market opened on Thursdays, Fridays, and Saturdays, giving clothiers the first part of the week to get to town. If a clothier didn't sell all his merchandise before returning home, he could store it or ask another clothier to sell it for him.

Like the Champagne fairs, in other words, Blackwell Hall was designed as a destination for people traveling to sell their textiles on a specific schedule. It, too, evolved into something more convenient. Instead of going back and forth, a clothier could contract with a London factor to sell his textiles on commission. In the early days, factors came from many different backgrounds. One could, it was said, choose "a Factor from almost any profession, an Oyl-man, a cloth-drawer, a Tobacconist, etc." At the end of the seventeenth century, around three dozen factors handled the business at Blackwell Hall, each representing numerous clothiers. A 1678 law officially recognized their function.[47]

A factor maintained an inventory of his clients' cloth, keeping a few hundred pieces in stock at a time. When a fabric wholesaler, known as a *draper*, or an exporter expressed interest in a particular type of textile, the factor would send out samples. The buyer could purchase the cloth in inventory or place a custom order. Buying existing cloth was faster, of course, and the merchant knew exactly what he was getting. Because of differences in dye lots or spinning, even a seemingly exact copy might not turn out the same. But custom production, often hurried, was nonetheless common.

To reduce the chances of unsold inventory, factors carefully monitored market trends. "By conversation in the office, in Blackwell Hall, and in the coffee houses, and by watching the trend of fashion," writes historian Conrad Gill, who examined the correspondence between a London firm and its clients in the West Country,

> factors used to gather information about the probable course of demand, and made forecasts which they passed on to the clothiers. Makers of cassimeres [twills used in suiting], for instance, were told in 1795 that there was a demand for white pieces, not of the natural colour of wool, but carefully bleached and dressed. . . . The firm which was advised to make the bleached cassimeres was told at the same time that cloths of various other colours could be sold—lemon buff, good and varied drabs, a small number of scarlets; and a few days later dark blue supers [the highest-quality woolens] were mentioned.

Factors often sent their clients detailed proposals for patterns they expected to sell well.

Along with long-distance representation and market intelligence, factors provided quality control, addressing the problems of poor workmanship and outright fraud that often plagued textile markets. To save on yarn, weavers might reduce cloth dimensions or use a higher density weft at the beginning of a bolt, where it was readily visible, while skimping further along.[48] Or they might try to hide poorly spun yarn with aggressive fulling, in which wool fabric is shrunk so the fibers compress. Up until 1699, a government inspector, known as the *aulnager*, certified the size and quality of woolen cloth, but the inspections tended to be perfunctory and focused primarily on the cloth's dimensions. The aulnager's main function seemed to be collecting a tax on each piece of cloth.

With their reputations on the line, writes Gill, "factors did more than aulnagers had ever done, for they strove constantly to ensure that the pieces which passed through their warehouses should be not only of the right dimensions, but also as free as possible from faults of any kind." An individual clothier might himself develop a reputation for reliable cloth, and many did. But factors amplified the effect. While aggregating supplies from many sources, they dealt repeatedly with the same customers. They were less likely to be tempted by the short-term gain of selling inferior goods to buyers they might not see again. Reliable quality brought rewards.

But maintaining standards meant sometimes turning down material into which clothiers had sunk time and money, expecting a return. Cloth could be rejected for uneven color, stains, or minor holes; it might be too thin, too coarse, too dirty, or just "very bad." If the clothier was generally reliable, the factor would offer constructive criticism. When dealing with consistently poor goods, however, factors could be brutally blunt. Francis Hanson, one of the partners whose correspondence Gill reviewed, told a clothier to forget trying to sell his "infamous" merchandise in London and stick to the country, where expectations were lower.

To satisfy factors' quality demands, clothiers, in turn, had to impose consistent standards on their contractors. Was a clothier oppressing and defrauding a weaver, dyer, or spinner if he refused to pay for material he found flawed? To workers living hand to mouth, it seemed so.

Over time, factors took on additional roles. They began to purchase wool and sell it to clothiers. They bought cloth on behalf of foreign merchants, who were forbidden entrance to Blackwell Hall. When demand was high, they served as clothiers themselves, much to the annoyance of their clients. And they provided credit. They lent to merchants buying cloth and to clothiers buying wool. They advanced clothiers money against the sales of their cloth.[49]

All these functions made the cloth market run more smoothly. But clothiers chafed at their dependence on the middlemen. "I have heard many of them say, that if a Legislature would deliver them from this insupportable Yoke," declared Trowbridge, "they would oblige themselves both to increase their Wages, and lower the Price of their Goods."[50]

Disgruntled clothiers imagined that factors were settling for excessively low prices, rejecting fabric without reason, and making money without effort. They felt the pinch of their loan payments and resented the factors for buying wool to sell them at a profit. They forgot the services that had led them to rely on factors in the first place: the convenience, working capital, market intelligence, quality control, and customer connections. Especially in hard times, it was easy to perceive the cost of the middlemen's work, in commissions and interest, but not its benefits.

▨▨▨▨

It is the antebellum South, a few years before the Civil War. Smitten with a young woman named Babette Newgass, Mayer Lehman calls on her father to ask for her hand in marriage. Mayer is the youngest of three Jewish brothers who have immigrated from Bavaria and set up shop in Montgomery, Alabama. The well-to-do Mr. Newgass appears skeptical about his would-be son-in-law's prospects.

MR. NEWGASS: Seeing as how you present yourself, I'd like to know, young man, what exactly it is you Lehmans do in your store.

MAYER: We used to sell fabrics, Mr. Newgass, but now we don't anymore.

MR. NEWGASS: If you no longer sell fabric, what use do you have for a store?

MAYER: Oh, we're still selling, Mr. Newgass.

MR. NEWGASS: What are you selling?

MAYER: We sell cotton, Mr. Newgass.

MR. NEWGASS: Isn't cotton a fabric?

MAYER: Not when we sell it, Mr. Newgass. When we sell it, it's raw.

MR. NEWGASS: Who buys it from you?

MAYER: Men who turn it into fabric, Mr. Newgass. We're in the middle, right in the middle.

MR. NEWGASS: What kind of job is that?

MAYER: Something that doesn't yet exist. Something that we invented.

MR. NEWGASS: What is it?

MAYER: We're . . . middlemen.[51]

The characters are real historical figures. The scene is imagined. It comes from *The Lehman Trilogy*, an epic, five-hour work by the Italian playwright Stefano Massini, condensed to a mere three hours in English. The New Yorkers who packed the sold-out run at the Park Avenue Armory in April 2019 got a rare reminder of the textile origins of Lehman Brothers, the fabled investment bank whose collapse in 2008 came to symbolize Wall Street's failed promises.

Based in history, *The Lehman Trilogy* is also fiction, just as Shakespeare's *Henry V* and *Julius Caesar* are both histories and fictions. When a New York friend who saw the opening-night show told me that the Lehmans had invented the middleman, I thought she must have misunderstood.

After all, Pūsu-kēn and his fellow traders were middlemen thirty-nine hundred years before a Lehman got off the boat from Bavaria. Nor were the brothers unique in the antebellum South. Like its predecessors in wool, silk, and linen, the nineteenth-century cotton trade relied on middlemen—first known as factors and later, reflecting organizational changes wrought by the railroad and telegraph, as brokers. (After the Civil War, my own ancestors went into the business, operating from Atlanta and New York.) The middleman role was a familiar one, imported from the Old Country. The Lehmans certainly didn't invent it.

Brokers provided cotton growers with working capital, crop transportation, and networks of buyers. They gauged cotton quality and anticipated prices. Before the Civil War, they also supplied goods. "Whether the planter desired a set of books for his library or shoes for his slaves, several bottles of imported brandy or a barrel of western pork, he had only to ask his factor and the goods would be purchased and sent to the plantation," writes a historian.[52] After the war, cotton brokers became increasingly sophisticated. In the 1870s, they established exchanges in New York and New Orleans to track prices and facilitate the trading of futures contracts as a hedge against price fluctuations.

Mayer's claim to invention reflects artistic license. With the conversation, Massini captures the puzzlement and apprehension that the middleman role inspires. What do these people do? What value do they add? *What kind of job is that?*

Edgar Degas's uncle was a New Orleans cotton broker in 1873, when the artist painted A Cotton Office in New Orleans, *the first of his works to be purchased by a museum. (Wikimedia)*

Earlier, the playwright invents a crisis—a devastating fire that destroys Montgomery's cotton crop—to explain how shopkeepers became cotton merchants. To provide the seeds and tools needed to replant, the Lehmans accept pledges of a third of the next harvest. Here, in brief, is what middlemen do. They build the economic bridge between today and tomorrow, and they charge a toll.

The audience watches as generations of Lehmans invest in coffee and cigarettes, railroads and airlines, radio and movies, and finally computers. "The Lehman Brothers history," says Massini, "is not only the story of a family and of a bank. It is the history of our last century." Cotton is quickly forgotten, more quickly in the play than it was in real life.

The Lehmans help found the New York Stock Exchange, a social technology described anxiously as "a temple of words." The stock exchange doesn't deal in actual commodities, grumbles Mayer, only words: "There's no iron, there's no fabric, there's no coal, there's nothing there." The exchange descends from Fibonacci's arithmetic, which taught the West to record its business with symbols on a page—mere ink, suspiciously intangible.

The Lehman Trilogy is not a morality tale. It is ambivalent, perceiving both possibilities and dangers in the alchemy of finance. Its characters are neither angels nor devils, but human beings. "There was a moment in my life," recalls Massini, recounting the play's inspiration, "when I found that people in Italy, in Europe, perhaps in the United States too, were *hating* the economists, the banks, the great finance. In that moment I thought that I need to write the history, not of a terrible bank with terrible men, but the incredible history of the very human foundation of a bank. I think the story of Lehman is the very, very human history of the foundation of a great empire."[53]

An American would have written a polemic about greed and disaster. Many US critics saw *The Lehman Trilogy* that way—"a religious parable of reckoning," one called it—or condemned it for slighting the sins of slavery.[54] But Massini comes from a place familiar with the rise and fall of great fortunes, the ambiguous lives and lasting legacies of merchant bankers, the complexities of history, and the necessity of credit. He is a Florentine.

Chapter Six

CONSUMERS

Nowadays the very servant girls dress in silk
gauze, and the singsong girls look down on
brocaded silks and embroidered gowns.

—Tian Yiheng, *Liuqing ri zha*, 1573

IN A HAND SCROLL PAINTED around 1145, a Chinese silk weaver sits at her large floor loom, a portrait of concentration as she beats the weft into place. Her lips pursed, she depresses a treadle with her bare foot. Her left hand readies the shuttle for its next pass. It takes three days of steady work to weave a single bolt of silk, about thirteen yards long, enough to outfit two women in blouses and trousers. But the weaver herself does not wear silk.[1]

Accompanying the picture is a poem addressed to those who do:

Working with diligence, repeatedly passing the shuttle,
Starting across, again returning [with the shuttle] makes a shed.
Sending this poem to the couple in damask silk,
[They] should think of the one who wears coarse hemp.

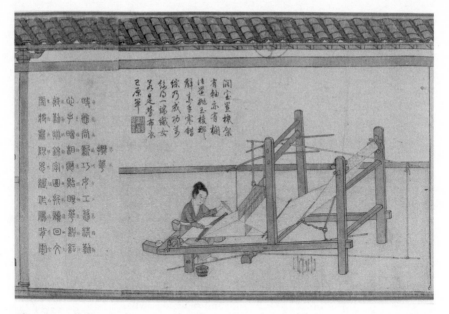

She weaves silk for taxes but wears coarse hemp. From the scroll Pictures of Tilling and Weaving *after Lou Shu, attributed to Cheng Qi, active mid- to late thirteenth century (Freer Gallery of Art, Smithsonian Institution, Washington, DC: Purchase—Charles Lang Freer Endowment, F1954.20)*

In the scroll, titled "Pictures of Tilling and Weaving," a local magistrate named Lou Shu immortalizes in meticulous detail twenty-four different stages of sericulture, each with a poem capturing the emotions and experiences of rural life. In its day, the scroll was a moral and political work, aimed at influencing the powerful. "The agrarian laborers are presented as self-sufficient," writes an art historian, "and their welfare is posited as the justification for government." Lou's work encouraged officials to respect farmers' humanity and competence and to use their taxes wisely.[2]

It was a worthy goal. But as a historical artifact the scroll perpetuates a common bias. Producers attract our interest and sympathy. Consumers are denigrated or forgotten. Yet they are at least as important.

Without consumer desire, the story of textiles is incomprehensible and incomplete. The labor of spinners and weavers, the ingenuity of breeders, mechanics, and dye chemists, and the risk-taking ventures of merchants are not ends in themselves. They exist to serve the users of cloth. Those consumers include rulers demanding tribute, armies clothed and equipped with textiles, priests and sanctuaries draped in donations, and, of course, customers buying cloth in markets—both open and illicit.

The drive to acquire new fabrics is a surprisingly potent force. Whether purchasing cloth, making it for themselves, or seizing it from others, textile consumers defy expectations. They start wars and break laws, overturn hierarchies and flout traditions. Their shifting tastes reorder wealth and power, enrich upstarts, and leave former winners bereft. Their choices challenge static notions of authenticity and identity. Textile consumers change the world.

𝌀𝌀𝌀𝌀

For the Southern Song government (1127–1279 CE) of Lou's day, silk was essential to maintaining power and keeping the peace.[3] The emperor's regime used the precious fabric to buy off the rival kingdoms threatening its borders, to outfit its expanding army, to reward loyal officials, and to bestow gifts on commoners. Each year, the Chinese state bought four million bolts of silk. It collected more than three million additional bolts in taxes. Behind those taxes was the labor of countless peasants clad in modest hemp.

In the final panel of Lou's scroll, three women measure out bolts of fabric, folding them to go into a basket for the tax collector. The accompanying poem pronounces the sericulturalists' labors worth the effort. The empire is putting the silk to good use, it declares, not siphoning it off for elite enjoyment.

> *Tax officials transport the silk for border defense,*
> *All that hard work! But no need for pity.*
> *This is a greater victory for the fine woven silk of the Han;*
> *When before, if the silk were stained by the rouge*
> *[of a courtesan], it was never worn again.*

With its invocation of the extravagant courtesan—by implication the mistress of a corrupt official—and the couple in damask silk, Lou's moralizing hints at another source of silk demand: the burgeoning consumer market.

During the Song, China experienced its own commercial revolution (complete with bills of exchange known as *feiqian*, or "flying money"). Textile markets flourished. Public and private silk consumption amounted to as many as one hundred million bolts a year. About twenty million came from urban artisans specializing in luxury fabrics, while the rest were made on simpler looms in the countryside. "Textile manufacture, previously confined mostly to domestic needs and tax payments, became reoriented to production for the market," writes a historian. Taking advantage of high prices, farm families turned to full-time silk production.[4]

In the cities, fabric shops flourished. In the capital of Hangzhou, writes a textile scholar, specialty stores included "Chen's silk shop in the city's west, Xu's embroidery shop near Water Lane, the raw silk shop underneath the Water Lane bridge, Gu's silk shop in the Qinghe district, and hemp and ramie cloth shops near the Pingjin Bridge."[5]

Responding to the popularity of lightweight, intricately patterned silk *gauze* among the wealthiest buyers, rural producers created a more affordable alternative. They designed airy plain weave fabrics known as *open tabby* that didn't require the skill and special equipment of gauze, which is created by twisting pairs of warp threads together and inserting the weft through the twist. Among the novel motifs were "chestnuts," "fluffy loops," and "clear skies." This inventiveness found eager buyers in the middle market between the poem's "damask silk" and "coarse hemp."

China's silk expertise attracted foreign attention as well, not all of it satisfied by tribute and trade. The best surviving version of Lou's scroll is not the Southern Song original but a copy dating to the Yuan dynasty, when warriors from the steppes ruled China—and much of the world.

◇◇◇◇

Beginning with Genghis Khan's unification of warring steppe tribes in 1206, the Mongols built the largest continental empire in history. By the close of the thirteenth century, their domain stretched from the Sea of Japan to the Danube. Descendants of Genghis Khan ruled China, Russia, and Iran.

The Mongols did not weave. Theirs was a nomadic culture of furs and *felt*, which is made by using friction to mat together wet animal fibers. But they treasured woven textiles, and their desire for fine fabrics motivated many of their conquests. "The common thread running through all the inventories of plunder is rare and colorful textiles, tenting, and clothing," writes historian Thomas Allsen. To furnish his capital with cloth, Genghis Khan deported weavers to Karakorum from conquered territories.[6]

Blending the indigenous with the imported, Mongol rulers received visitors in great tents whose exterior walls of white felt were lined with silk brocaded in gold. Known as *nasīj*, this style of fabric originated in Muslim lands far to the west of Mongolia. But it became so identified with the Mongols that Europeans referred to it as "Tartar cloth" and "cloth of Tartary," using a general term for people of the steppes.[7]

"The significance of luxury textiles in the Mongol Empire can hardly be overstated," observes an art historian specializing in Asian textiles.

Plunder, trade, diplomacy, ceremony, and tribute and taxation were occasions for the acquisition, distribution, and display of cloth—especially opulent silk textiles woven with gold—activities that were frequently public and symbolic of Mongol political power. Luxury textiles found many uses:

The desire for fine fabrics like the cloth of gold in this caftan spurred Mongol conquests. Most of the areas that were originally gold on this garment have turned brown over time. (The David Collection, Copenhagen, 23/2004. Photo: Pernille Klemp)

in garments and personal accessories, horse and elephant trappings, tent and palace hangings, cushions and canopies, works of religious art, and even imperial portraiture.[8]

When the Mongols invaded Afghanistan in 1221, the city of Herat was among their greatest prizes. A weaving center famous for its cloth of gold, Herat surrendered without a fight, sparing its inhabitants the slaughter visited upon those who resisted. (That fate would come the following year, after a revolt against the occupation.) Along with the usual plunder, the Mongols seized a particularly valuable textile treasure: as many as a thousand skilled weavers.

They transported these captives more than fifteen hundred miles across central Asia to the Uyghur capital of Beshbalik, in what is now the Xinjiang region of northwest China, closer to the Mongol heartland. (The Uyghurs were the first foreign kingdom to surrender to Mongol rule and were themselves known for their silk tapestry.) With their involuntarily imported talent, the Mongols established a weaving colony to produce *nasīj*. Soon Beshbalik, historically a town of Buddhists and Nestorian Christians, had a flourishing Muslim community, seeded by the displaced weavers of Herat.

As the Mongols gradually conquered China, with Kublai Khan establishing the Yuan dynasty in 1279, they used forced migration to create new textile centers. Because China had its own vibrant silk tradition, these relocations did more than ensure a convenient supply of cloth. In what seems to have been a deliberate policy, Mongol workshops encouraged the exchange of techniques and motifs.

As they set up workshops to satisfy their textile cravings, the Mongols mixed together artisans from different places. They transported weavers from Samarkand, in what is now Uzbekistan, to the city of Xunmalin near modern Beijing; they also transferred Chinese weavers to Samarkand. They sent three hundred artisans from western conquests and another three hundred from northern China to a new settlement in Hongzhou, west of Beijing.

"Under Mongolian auspices large numbers of West Asian weavers and textile workers, not just the products of their looms, were sent east and became permanent residents of China," writes Allsen. "This may not be

This Mongol textile combines Chinese and Iranian motifs and techniques. (Cleveland Museum of Art)

entirely unprecedented but surely the scale on which these forced resettlements were undertaken was extraordinary." Brutal and inhumane, the relocations nonetheless "created unparalleled opportunities for technical and artistic exchange."[9] The result, and perhaps the goal, was a profusion of novel patterns.

A Mongol cloth of gold owned by the Cleveland Museum of Art illustrates the hybrid designs that emerged from imperial workshops. It combines Iranian motifs of griffins and winged lions with Chinese cloud patterns on the lions' wings. The gold threads creating the design against a base of dark-brown silk are made by affixing the metal to a paper substrate, a Chinese technique, while the weave structure, known as *lampas*, originated in Iran.

"Textiles from the Mongol domain of this period defy any attempt at definition," writes a textile historian.

With the flourishing trade of the Mongol Empire, designs migrated across cultural boundaries, combining a mixture of traditional Chinese motifs, Middle Eastern elements and local Central Asian repertoires, and for a brief period there was an international decorative repertoire in Chinese, Middle Eastern, Mamluk and Lucchese silks. Entire colonies of skilled artisans of mixed ethnicity were created, which facilitated a kind of hybrid development in textile art and its technology—bewildering textile historians today.[10]

As diplomats and merchants imported hybrid textiles, the creative ferment reached beyond Mongol territory to influence European styles. "In Italy," write two art historians, "the impact of their exotic designs triggered the most imaginative chapter in the history of European silk weaving."[11]

For the same reason they looted cities and captured artisans, the Mongols encouraged commerce. They wanted stuff—textiles in particular. "The close connection between their interest in long-distance trade and the desire for luxury textiles is frequently mirrored in the sources," Allsen observes. "In one of his maxims, Chinggis Qan [Genghis Khan] extols the virtues of merchants who 'come with garments of gold brocade' and even proclaims them as role models for his military officers!"[12]

When the wars of conquest finally ended in 1260, the ensuing Pax Mongolica created a vast expanse for peaceful exchange, as the Mongols turned their former military routes into protected commercial arteries. Along with silk, Mongol trade routes brought new ideas and technologies to Europe from the East, including gunpowder, the compass, printing, and papermaking. The Black Death came, too. The result of the Mongols' ruthless pursuit of textiles was a synthesis of cultures, motifs, and weaving techniques—and a world forever changed.

〰〰〰〰

In 1368 Zhu Yuanzhang ascended the throne as the Hongwu emperor, founder of China's Ming dynasty. Born a peasant, Zhu had commanded an army fighting to overthrow the Mongol Yuan dynasty, eventually triumphing over both the old regime and his rival rebels. Once in power, he sought to restore what he viewed as traditional Han order after nearly a century of barbarian rule.

One of his first acts was to establish a dress code. It banned Mongol styles and dictated standards for each rank of government officials, distinguishing them from each other and from ordinary people. Other rules reinforced the neo-Confucian hierarchy of commoners: scholars, farmers, artisans, and merchants. The code regulated clothing materials, colors, sleeve lengths, headgear, jewelry, and embroidery motifs. The goal, the emperor declared, was "to make the honored and the mean distinct and to make status and authority explicit."[13]

Rather than restricting garment styles, most of the rules governed who could use what types of textiles. Commoners were forbidden to wear silk, satin, or brocade. The stricture was relaxed for farmers in 1381, allowing them silk, gauze, and cotton. But if any member of the family engaged in commerce, no one could wear silk. Merchants, while useful, were to be kept in their place.

"The basic function of the Ming clothing system was to impose state control over the whole society," writes a historian. "If the whole society was shaped exactly by the regulations and continued these regulations forever, it would be a model Confucian society, stable and stratified." That was the theory, at least.[14]

For the nearly three centuries of Ming rule, the regulations did remain largely unchanged.[15] From time to time, penalties for violations were increased. The society did not, however, remain stable. The rituals central to Confucian order fell out of use or took on discordant elements, as when funerals included actors, musicians, and prostitutes as entertainment. Daoist and Buddhist practices seeped into Confucian culture. As commerce flourished, merchant families grew wealthy and prominent, sometimes assuming aristocratic status.

And people didn't follow the rules. "Archaeological evidence from the tombs of Ming princes shows that Mongol styles of dress persisted well into the sixteenth century," writes historian BuYun Chen, "thus revealing both the limits of Zhu Yuanzhang's sartorial code and suggesting, more seriously, the failure of his efforts to eradicate the legacy of the Mongol Yuan."[16]

As time passed and commerce grew, violations increased. Wealthy commoners dressed in fabrics and styles supposedly reserved for nobler classes. They scorned plain silks and adopted forbidden brocades. They wore off-limit colors, including dark blue and scarlet. They sported gold embroidery. They bought hats and robes supposedly restricted to court officials. "Customs have changed from generation to generation," complained a Ming scholar, writing in the late sixteenth century. "All people tend to respect and admire wealth and luxury, competing for them without considering the bans of the government."

Nor were commoners the only offenders. Officials and their families dressed above their station. The sons of nobles, themselves in the lowly eighth rank, habitually donned the dress reserved for their high-ranking fathers. "They wear dark brown hats and robes patterned with *qilin*," a dragon-like creature with cloven hoofs, "tied with golden ribbons, even when they live at home or have been dismissed from official positions," complained another Ming writer. Emperors themselves undermined the rules, he observed, bestowing robes on favorites without regard to whether their status merited the design.[17]

Despite their contempt for the law, Ming consumers paradoxically re-affirmed the hierarchy it was meant to enforce. They didn't crave *qilin* robes because they were more beautiful or luxurious than similar garments with different motifs. They wanted them because of the association with high-ranking court officials. *Sumptuary law* defined what was desirable—and the most desirable goods were symbols of imperial status. As a result, argues Chen, "Imitation did not necessarily diminish court power. Conspicuous competition to put on the raiment of state-sanctioned power reaffirmed the emperor's place at the centre of the empire."[18]

The contrast with Edo Japan (1603–1868) is telling. There, the Tokugawa shogunate established its own Confucian-inspired hierarchy with sumptuary rules to match. (Low-level samurai replaced scholars as the top commoners in the Japanese ranking.) The laws were so continually flouted and revised that people mocked them as "three days laws."

Instead of aping their supposed betters, however, the urban artisans and merchants classified as lowly *chōnin*, or townspeople, invented new ways of embellishing and wearing textiles that skirted the restrictions and defined sophisticated taste. When the law decreed tie-dyed *shibori* patterns off-limits, they developed methods of hand-painting silk. Forbidden to appear in bright colors, well-dressed urbanites kept the exteriors of their clothes plain and hid the luxury in the linings, developing a sense of style called *iki* in which subtlety was paramount. "How better to sidestep the stiff samurai who forbids you to wear gold-embroidered figured silk," writes anthropologist Liza Darby,

> than to wear a dark-blue-striped kosode of homely wild silk—but line it in gorgeous yellow patterned crepe? Or commission the lining of your plain jacket to be painted by one of the foremost artists of the city? One got the satisfaction not only of complying with the law but also of one-upping its snobbish perpetrators. Relentless arbiters of style, the townspeople turned the fashion tables back in their favor by disdaining the gorgeous ostentation now denied them. Let the samurai and prostitutes cling to colorful brocades. Anyone with taste would turn to the subtler details marking a person as iki.[19]

Here, sumptuary law didn't set the standards of fashion. Wealthy merchants and kabuki stars did. In China, where high scores on exams could turn a peasant into a government official, ambition still focused on the court. The goal was to climb a static hierarchy, and clothing choices, however forbidden, reflected that ambition. In Japan, commoners didn't aspire to be samurai. They valued an urban life of art, pleasure, and fashionable innovation. But in both places, people used textiles to express who they wanted to be.

𝍠𝍠𝍠𝍠

As Zhu Yuanzhang was establishing Ming rule, at the other end of what would someday be known as the Silk Roads, Italy's mercantile republics were adopting their own restrictions on textiles, clothing, and adornment. From 1300 to 1500, Italian city-states enacted more than three hundred different sumptuary laws, "a greater number than in all other areas of Europe combined," notes a historian. Padova limited women "whether married or not and of whatever status and condition" to two silk dresses. Bologna fined those who wore gilded silver fasteners. Venice forbade trains and "French fashions." Florence even specified that corpses could be buried only in plain wool, possibly lined with linen. The grave was no place for finery.[20]

Playing a young dandy, kabuki actor Ichikawa Yaozo III wears a dark kimono with a vivid red lining, complying with sumptuary laws while demonstrating iki *style. Print by Torii Kiyonaga, 1784. In Domenico Ghirlandaio's 1488 portrait, Giovanna Tornabuoni wears the figured brocade, floral embellishments, and crisscrossed stripes forbidden by Florentine sumptuary laws. (Metropolitan Museum of Art; Wikimedia)*

In city-states run by merchants, the rules were less concerned with maintaining social hierarchies than with curbing extravagance in general. Increasingly lavish displays might have offended the ascetic Christianity preached by Franciscan friars and the modesty and thrift valued by traditional merchants. But the paramount goal of sumptuary regulations had nothing to do with these traditions. It was financial self-discipline.

The laws sought to restrain the competitive pressure to spend ever greater sums on jewelry, textiles, and public celebrations. As worried about their household budgets as about the common good, ruling families hoped to slow the arms race of conspicuous consumption. Sumptuary laws gave them an excuse to say no, especially to their wives and daughters. (In Florence the rules were enforced by the tellingly titled *Ufficiale delle donne*, literally the "officials of women.")

Unlike the Ming, Italian city-states constantly revised the rules, trying without much success to get their citizens to comply. Analyzing Florentine sumptuary laws from the close of the thirteenth century to the end of the republic in 1532, historian Ronald Rainey found authorities repeatedly reiterating and revising the restrictions, to little avail. "Given the frequent enactment of sumptuary laws in the fourteenth century," he writes, "it is apparent that the commune's dress regulations were not being observed to the lawmakers' satisfaction."[21]

Florentine laws adopted in the early 1320s forbade women from owning more than four outfits appropriate for wearing in public. Of these, only one could be made of either *sciamito*, a costly silk, or *scarlatta*, a wool dyed with expensive kermes red. Then, in 1330, the city banned new *sciamiti* dresses altogether, requiring women who already owned them to register their garments with the city. In 1356, authorities outlawed even those exceptions, permitting only plain silk. Any woman wearing more elaborate textiles was subject to a stiff fine.

Laws changed to close loopholes and to adjust to fluctuating fashion. The 1320s law prohibited anyone, male or female, from wearing clothes decorated with images of "trees, flowers, animals, birds or any other figure, whether these figures were sewn on, cut into, or attached in any other way to the garment." A 1330 revision added painted figures to the list. It also banned sewing on stripes or crisscrossed materials to decorate women's dresses.[22]

Italian sumptuary laws may have deterred some extravagance, but they certainly didn't squelch it altogether. They simply encouraged stealth and fashionable workarounds—new styles that skirted restrictions. Hence the need to revise the law to ban silk stripes and painted figures.[23]

In one of his tales of Florentine life, the fourteenth-century writer Franco Sacchetti, who served as a sumptuary law official, captures the prevailing attitude. Hired to enforce the laws, a judge named Amerigo seems to be falling down on the job. Florence's women walk the streets in forbidden finery, yet he has charged no one with violations.

It isn't his fault, declares Amerigo. The women are simply too good at arguing the law. Stopped for wearing illegal embroidery on her hat, one alleged offender unpinned the decorative border and declared it a wreath. Another, questioned for wearing too many buttons, said the silver balls weren't buttons but beads. They had no matching buttonholes. Stumped by such logic, Amerigo says, he can't arrest the women. His bosses agree: "All the officers advised Messer Amerigo to do the best he could and to leave the rest alone." Sacchetti ends his story with a popular saying: "What woman wants the Lord wants, and what the Lord wants comes to pass."[24]

Someone breaking the sumptuary laws in Ming China risked corporal punishment, penal servitude, and confiscated goods. In Italy, the penalty was generally a fine. Dress restrictions served a fiscal purpose, filling city coffers.

Along with fines, the laws also generated fees. When new rules went into effect, a city usually offered citizens a way to keep their now-forbidden clothes: report the offending garment, pay a fee, and get a seal marking it as permitted. After Bologna enacted a new statute in 1401, more than two hundred garments were registered, generating at least a thousand lire in fines. (By way of comparison, a clerk earned a salary of sixty lire a year.) One woman bought permission to keep her green wool coat with forest imagery of deer, birds, and trees embroidered in gold. Another paid for five garments, including a coat of striped red wool and silver stars in a running wave pattern. A third registered a velvet dress adorned with gilded and scarlet leaves. "Fines and seal tagging became a sort of tax collection," observes a historian, suggesting "that fiscal motivations were one of the most powerful drivers of the policies adopted to regulate luxuries and appearances."[25]

Scrambling for revenue, Florence went a step further, turning its fines into de facto licenses. An annual fee, or *gabella*, could buy an exemption from an irritating restriction. Under the 1373 rules, fifty gold florins— enough for the commune to pay a crossbowman for fifteen months—gave a woman the right to wear woolen dresses embellished with silk patterns. For twenty-five florins, a married woman could decorate her hemlines, a privilege otherwise restricted to the unmarried. Ten florins let a man wear *pannos curtos* (literally "short cloths"), revealing his legs above the middle of his thighs when standing. For the same price, a woman could sport silk-covered buttons.

The list of exemptions available for a price was nearly as long as the list of prohibitions. "So extensive were these purchasable exemptions, in fact," writes Rainey, "that few items forbidden by earlier regulations remained altogether prohibited to women who could afford to pay the required taxes." The predictable result, he observes, "was to foster indifference among the Florentines to the regulation of conspicuous consumption."[26]

Despite occasional upwellings of ascetic zeal, most notably the Florentine friar Girolamo Savonarola's fiery sermons against luxury, the commercial cities of Italy lacked the conviction to seriously regulate finery or restrict it to a narrow few. In their hearts, their citizens believed that beautifully made things were good and brought honor to the wearer and the city. Even a gold-embroidered dress could point toward the divine.

Just as contemplating the "infinite works of nature" led some to perceive the greatness of God, wrote a Milanese defender of that city's traditional "freedom to dress" against Spanish-imposed sumptuary laws, so others

> contemplating the marvels of art, raise themselves in some way to a consideration of God's great wisdom, who infuses such knowledge into men, thereby comprehending in some way the great bounty of the self same God who, through His benignity, bestows ingenuity and industry on them; so they also glimpse the boundless and unintelligible Majesty of the self same God in Heaven upon seeing the majesty that rich garments and accessories confer upon earth.[27]

Places of commerce and industry, Italian cities knew their greatness depended on craftsmanship and consumer pleasure. While trying to restrain their acquisitive impulses through regulation, their citizens found honor in the creation and display of artistry of all sorts, including luxurious textiles and apparel. What the consumer wants tends to come to pass.

༺༻

On a shopping trip to the butcher's, young Miss la Genne wore her new, form-fitting jacket, a stylish cotton print with large brown flowers and red stripes on a white background. It got her arrested.

Another young woman stood in the door of her boss's wine shop sporting a similar jacket with red flowers. She too was arrested.

So were Madame de Ville, the lady Coulange, and Madame Boite. Through the windows of their homes, law enforcement authorities spotted these unlucky women in clothing with red flowers printed on white. They were all busted for possession.[28]

It was Paris in 1730, and the printed cotton fabrics known as *toiles peintes* or *indiennes*—in English, calicoes, chintzes, and muslins—had been illegal since 1686. Every few years the authorities would reiterate and tweak the law, but the fashion refused to die. Frustrated by rampant smuggling and ubiquitous scofflaws, in 1726 the government increased penalties for traffickers and anyone helping them. Offenders could be sentenced to years pulling oars in the navy's galleys, with serious traffickers put to death. Local authorities were given the power to detain without trial anyone who merely wore the forbidden fabrics or decorated their homes with them.

"The exasperation of the lawmakers, after forty years of successive edicts and ordinances which had been largely ignored, flouted or circumvented on a wholesale basis, can be sensed in this law," writes fashion historian Gillian Crosby. Its main effect was a crackdown on consumers, with a spike in arrests for simple possession. "Impotent at stopping the cross-border trade,

From 1686 to 1759, owning a printed chair seat like this one in France could get you sent to prison. (Metropolitan Museum of Art)

printing or the peddling of goods," Crosby writes, "government officials concentrated on making an example of individual wearers, in an attempt to halt the fashion." They failed.[29]

In the annals of prohibition, the French war on cotton prints is one of the strangest and most extreme chapters. The ban was not a sumptuary law but a draconian form of economic protectionism designed to insulate established industries from consumer tastes. The original 1686 prohibition explained:

> The King has been informed that the great quantity of cotton fabrics painted in the Indies or counterfeited in the Kingdom . . . have not only given rise to the conveyance of many millions outside the Kingdom, but have also caused the reduction of Manufactures long established in France for Stuffs of Silk, Wool, Linen, Hemp, & provoked at the same time the ruin & desertion of Workers who, no longer finding employment nor subsistence for their families due to the cessation of their work, have left the Kingdom.[30]

Other European countries, including England, also banned calico imports, but French policy was the most drastic. It did more than block imported prints. It also prohibited plain cottons from abroad. And it forbade domestic printing, even on French-made cloth. It was not just anti-foreign; it was anti-cotton and anti-print. England, by contrast, fostered a domestic industry of printing on fustian, which used linen warp threads and cotton weft.[31] The French prohibition lasted the longest in Europe, stretching for seventy-three years. It never worked. Consumers loved calicoes and refused to give them up.

Introduced by Portuguese traders in the sixteenth century, the Indian fabrics were unlike anything Europeans had ever seen. The blues and reds were spectacular and, thanks to dyeing skills honed over centuries, the colors survived frequent washing. The cotton fabric was soft and lightweight, ideal for summer clothes and more comfortable than linen as underwear. Prints themselves were largely new to Europe, providing an irresistible cornucopia of pictorial designs without the expense of drawloom weaving.

Savvy Indian producers modified their patterns to suit local tastes, as they'd long done for customers in East Asia. The most important adaptation was printing or painting the colorful designs on white backgrounds rather than using white figures on blue or red. Blocking large expanses of cloth from absorbing dye required new techniques. So "consumers in Europe did not just remodel products," observes a historian, "but also shaped the innovative technologies used to produce them." The resulting textile patterns were hybrids, blending designs from Europe and Asia to produce motifs that were at once familiar and fashionably exotic.[32]

Indiennes weren't exclusively luxury goods. The prints offered options for every income. An aristocratic lady might wear a finely painted skirt to court while, for less than a day's wages, a servant girl could spruce up a drab ensemble with a floral kerchief. "The trick to their success," writes historian Felicia Gottmann, "was that they came in a vast range of qualities, from the finest hand-painted chintzes to the cheapest block-printed or dyed calicoes, and thus could furnish the summer houses of aristocrats, clothe poor labourers, and provide the bourgeoisie with cheaper alternatives to high-quality French silks."[33] By the mid-seventeenth century, *indiennes* were everywhere, on everyone.

The cotton cloth's spectacular success triggered political opposition from silk, linen, and wool manufacturers, whose voices counted more at Versailles than those of mere consumers. Industry representatives persuaded the government to make the upstart fabrics illegal. But, from the beginning, smugglers exploited every conceivable loophole.

The regime didn't want to completely deprive the government-controlled French East India Company of its European market. So the law permitted auctions of *indiennes* supposedly destined for buyers abroad. The textile auctions attracted bidders stocking up on prints to trade for slaves in West Africa, which was legal, and those planning to sell fabrics in the French West Indies, which was not. It was all but impossible to tell who was up to what.

Officially legitimate foreign buyers had suspicious motives. Many were from Switzerland and the Channel Islands, notorious sources of illicit textiles.[34] Foreigners bought calicoes at auction, took them home, and smuggled them back in. Forbidden fabrics came across the border from Holland and Savoy, where they were legal. They crept in through Avignon, which was

Calicoes allowed everyone from lords and ladies to maids and prostitutes, such as the "St. Giles's Beauty" in this eighteenth-century print, to enjoy colors, patterns, and comfortable cloth. (Courtesy of The Lewis Walpole Library, Yale University)

under papal control, and out of ships and warehouses in Marseille, where they were supposedly destined for re-export.

Anyone in France who wanted calico—and most everyone did—could get it. Worn by the country's most fashionable women within sight of its most powerful men, the prestige of *indiennes* never dimmed. Instead of building the kingdom's wealth, the ban turned countless citizens into outlaws.

It also stifled the development of a French fabric-printing industry, even as entrepreneurs in England, Holland, and Switzerland were devising successful printing techniques. Their prints weren't as fine as Indian textiles, but they were good enough for many customers, including French buyers.

Prohibition also had an intellectual consequence. Amid the period's Enlightenment ferment, the ban produced some of the earliest arguments for economic liberalism. "Long before the more famous debates about the liberalisation of the grain trade, about taxation, or even about the monopoly of

the French Indies Company, *philosophes* and Enlightenment political economists saw the calico debate as their first important battleground," writes
Gottmann.[35]

To the mercantilist argument that permitting calico production would
be good for French industry, economic liberals added a novel point. The law
was unjust, they argued, in penalizing the many for the benefit of the few.
What textile producers were demanding was barbaric. Wrote the Abbé An-
dré Morellet in a 1758 tract against the ban:

> Is it not strange that an otherwise respectable order of citizens solicits terri-
> ble punishments such as death and the galleys against Frenchmen, & does so
> for reasons of commercial interest? Will our descendants be able to believe
> that our nation was truly as enlightened and civilized as we now like to say
> when they read that in the middle of the eighteenth century a man in France
> was hanged for buying in Geneva at 22 sous what he was able to sell in
> Grenoble for 58?

The textile business, he reminded readers, was not the French nation,
but only a tiny fraction of it. "It was the cruelty of the *system* of repression,
not any single instance of it, that the author wished to emphasize," writes a
historian.[36]

Worn down by public resistance and intellectual arguments, and concerned about European rivals developing their own printing industries,
in the 1740s the regime granted a few entrepreneurs the right to print on
domestic fabrics, including cotton from French colonies. Once these enterprises produced acceptable prints, the clamor for legalization increased.
After all, even Jean-Baptiste Colbert, the father of French *dirigisme*, had
argued only for protecting new industries, not well-established ones like
Lyonnais silk.

The ban was lifted in 1759, giving anti-prohibitionists a partial victory.
The regime opted for a 25 percent duty, which kept smuggling profitable.
Once in the country, the duty-skirting fabrics could easily pass as legitimate. Nonetheless, and despite their late start, French entrepreneurs were
able to develop a successful calico industry, eventually perfecting the new

technology of copperplate printing, which drew on Europe's well-developed business of using engravings to illustrate books. Domestic cottons such as *toile de jouy*, featuring intricate vignettes inspired by Chinese porcelains, became as fashionable as exotic *indiennes*.[37] And French citizens were no longer imprisoned for donning flowery aprons, sitting on chintz upholstery, or bedecking their beds with *toiles peintes*.

⊠⊠⊠

No matter what people might imagine back in England, Richard Miles knew his customers weren't rubes. They wouldn't accept any old baubles a foreign merchant might offer. They were picky and brand conscious, and he needed to keep them happy.

So when he wrote home for new supplies, Miles was specific and blunt. Send a few lengths of blue cloth, he instructed, "no Green ones. A few Yellow I believe will sell for Gold." Customers, he continued, liked the light wool twills, known as *half says*, produced by a manufacturer named Knipe much better than his competitor's.

> I'm really sorry to tell you that Mr Kershaw's half says are by no means equal to [Knipe's] nor do I think if all the men in the kingdom were to attempt a manufacture of the kind, they cd. eclipse Knipe's; at least not in the eyes of the Black traders here, & it is them that are to be pleased.

It was 1777, and Miles was an officer in the Company of Merchants Trading to Africa, commanding a fort in what is now Ghana. On the side, he conducted his own private business, swapping imported goods for gold, ivory, and, most of all, slaves.

During his posting, from 1772 to 1780, Miles bought 2,218 enslaved Africans in 1,308 barter transactions. He traded primarily with the Fante residents of the coastal areas, who acted as middlemen for Asante slave-catchers selling captives from the interior. Since the Fante traded both for themselves and for their Asante suppliers, historian George Metcalf notes, "there is little doubt that the merchandise Miles bartered in this area comprised the goods most in demand throughout the whole of the territory inhabited by the Akan peoples," a group that includes the Asante and Fante. In exchange for slaves, they wanted textiles.

Analyzing the detailed records Miles kept of his trades, Metcalf found that cloth accounted for slightly more than half the value of goods bartered for slaves, with gold in second place at about 16 percent. Ignoring gold, which essentially functioned as currency, textiles rose to more than 60 percent. "In terms of the Akan consumer," Metcalf observes, "it is no exaggeration to say that textiles were what the trade was all about."[38]

Like the Mongols before them and the Europeans with whom they traded, the Fante and Asante showed few humanitarian scruples about the brutal cost of their textiles. Even before cotton conquered the American South, the slave trade was thoroughly entangled with cloth—driven by demand from West African consumers.[39]

In the area's hot climate, lightweight fabrics were what buyers most desired, with cottons making up about 60 percent of the textiles Miles bartered for slaves. "A beautiful *indienne* will always fetch more than another more costly cloth," observed a French writer, "either because the variety of colors is more to the taste of Negroes, or the lightness of the cloth is more suited to these hot climates." Often made specifically for the African market, such prints came to be known as "Guinea cloth." To get what they really wanted, Asante also traded European imports for a blue-and-white-striped cotton cloth known as *kyekye* from what is now the Ivory Coast. They preferred this soft, firm fabric to anything made with imported yarns.[40]

West Africans knew what they wanted, in other words, and it was often different from what Europeans were used to making. Woven patterns in indigo and white did best, following local customs. Just as Indian producers had adjusted to European tastes, so European textile makers sought to please their customers in Africa. To understand—and copy—what was likely to sell, English textile manufacturers instructed their agents to send back samples of indigenous cloth. "Although some of these attempts at imitations were more successful than others," a historian observes, "it is evident that West African tastes were having an impact on cotton textile production elsewhere."

West Africans appropriated that most characteristic of English textiles, scarlet wool, for their own purposes. In the Kingdom of Benin, on the coast of what is now Nigeria, it was a favorite fabric for royal attire, and the king allowed only those with his permission to wear it. Throughout the region, locals unraveled the fabric to reuse the yarn, ignoring its intended use. Combining the red wool with indigenous cotton or bast fibers, they wove brocades or added embroidery for ceremonial cloths. Local cellulose-based fibers didn't absorb color as well as protein-based wool.

"Dyed wool, especially wool dyed scarlet, would have been immediately appreciated for its exceptional luminosity—a visual power that was evidently seized upon by political and religious elites and deployed in their service," writes a historian. "What is most noteworthy about this case is that it shows how some specially chosen imported materials could be readily integrated with indigenous ones to transform significant ceremonial clothing traditions."[41] African consumers didn't simply take what was given. They imaginatively adapted foreign cloth to their own purposes, creating new textile hybrids.

You can see more recent hybrids on the streets of West and Central Africa today, in the brightly colored, mass-produced cottons known as *wax prints*. Called *ankara* in West Africa and *kitenge* in East Africa, the prints

Wax prints on display in a Ghanaian market. Now quintessentially African, the prints derive from Indonesian batiks by way of Dutch manufacturing. (iStockphoto)

were originally imitations of Javanese batiks intended for Indonesian customers. In the nineteenth century, Dutch manufacturers in the city of Haarlem perfected a roller process for printing with resin wax on both sides of the cloth. But the resin cracked during the process, leaving the distinctive lines that mark the fabrics to this day. Indonesians disliked the cracks, preferring hand-printed cloth, especially after batik makers developed less laborious techniques and cut prices. By the late nineteenth century, the Indonesian market had dried up.

Around 1890, the Scottish merchant Ebenezer Brown Fleming had the bright idea to try selling the machine-made fabrics on the African Gold Coast, now the country of Ghana. Perhaps he knew that locals liked batiks, which had been brought back as gifts by men serving in the Dutch army in Indonesia. Relying on hundreds of women traders to tell him what customers wanted, Brown Fleming didn't simply repeat Javanese patterns but tailored designs to African tastes. Because Africans were taller than Indonesians, he also changed the width of the fabric from thirty-six inches to forty-eight.

The colorful, highly polished fabrics proved a hit with upwardly mobile customers looking for better materials than the cheap English cottons previously available. Unlike Indonesians, African buyers liked the irregular lines produced by the cracking resin. "For them," observes an art historian, "these traits resonated with long-established and much-loved West African tie-dyeing and resist-printing techniques."[42]

As wax prints caught on, designs born in Europe took on decidedly local meanings. Cloth traders and customers gave patterns names derived from their own proverbs and life situations. "Names are the means by which consumers take possession of wax-print textiles, creating meaning that does not exist when the textiles are designed and produced," writes an art historian. So a print of curling stems titled "Leaf Trail" by its Dutch designer became "Good beads don't talk," a Ghanaian proverb counseling that truly admirable people don't brag. The pinwheel design its creator dubbed "Santana" became "Darling, don't turn your back on me" in the Ivory Coast.

"Speed Bird," a classic pattern of swallows in flight is "Money Flies" in some parts of Ghana and "Gossip Flies" in others. Ghana's innocuous

"Gramophone Record," named for its circular motifs, is called "Cow Dung" by polygynous customers in the Ivory Coast. (The name refers to a proverb suggesting that a household with multiple wives is not as peaceful as it seems: "Co-wife rivalry is like cow dung. The top is dry but the inside is sticky.") Women often wear a particular pattern to send a message, and the names are as essential to the cloth's value as the designs themselves. "Cloth traders and consumers alike agree," writes a curator who did fieldwork in Ghana, "that while women purchase cloth because 'it is beautiful,' they also buy it 'because it has a name.'"[43]

Although occasionally denounced as inauthentic, wax prints have become as thoroughly African as the indigo and white twill once called *serge de Nîmes* is American. "They are an intrinsic part of the social life of the people who use them," argues a textile scholar. Women collect and treasure uncut yardage, passing the cloth down to daughters and granddaughters. Limited-edition designs commemorate national celebrations and political campaigns. Wax prints play honored roles in weddings, funerals, baptisms, and baby namings. True wax prints, made the old-fashioned way at home and abroad, are luxury fabrics, but reproductions, including polyester versions from China, find their way into the poorest villages.

"These textiles are so completely absorbed into the patterns of daily life in many parts of Africa that they are everywhere but invisible," observes an art historian. "This is a major African art form, which is also a major European art form and a major Asian art form. It is, in short, complicated."[44] Textiles tend to be. The cultural authenticity of cloth arises not from the purity of its origins but from the ways in which individuals and groups turn textiles to their own purposes. Consumers, not producers, determine the meaning and value of textiles. Cloth is ubiquitous and adaptable, forever evolving in form and meaning. Trying to impose an external standard, heedless of consumers' beliefs and desires, is not merely futile but disrespectful and absurd.

⊠⊠⊠⊠

In September 2019, the US House of Representatives commemorated the four hundredth anniversary of the arrival of enslaved Africans in colonial America. To mark the occasion, members of the Congressional Black

At an event commemorating the four hundredth anniversary of the arrival of enslaved Africans on American soil, Speaker of the House Nancy Pelosi (right) and Representatives John Lewis and Sheila Jackson Lee wear stoles in a kente cloth design. (Getty Images)

Caucus and House leaders, including Speaker Nancy Pelosi, wore stoles in a pattern familiar from many an African American's graduation regalia. They were printed with alternating blocks, one with stripes of yellow, green, and red, the other black with a yellow motif in the center. The motif represents the Golden Stool, the symbol of kingship and power among the Asante. In their original environment, the blocks would be woven, not printed, and each four-inch strip would be one of twenty-four sewn together to form a single fabric, which is worn as a toga-like robe: Ghana's famous kente cloth.

For a thousand years or more, people in West Africa have created textiles by weaving strips a few inches wide and sewing them together side by side. But kente cloth emerged no earlier than the late eighteenth century. Its distinctive designs required colorful foreign thread, new loom technologies, and the cross-fertilization of Asante and Ewe (pronounced eh-weh) weaving practices. Once created, kente assumed shifting forms and meanings, at home and abroad, until the kingly robes of slave traders became the stoles honoring the heritage and achievements of those whose ancestors they enslaved.

For centuries, Asante weavers worked almost exclusively in white and blue. Aside from indigo, they had no intense colors for their cotton cloth. But they did trade. Across the Sahara from Libya and along the Atlantic coast from Europe came richly colored silks, exchanged for gold and slaves. By unraveling these fabrics, Asante weavers obtained bright threads with which to embellish their designs.

In his 1760 memoir, the Danish merchant Ludewig Ferdinand Rømer credits the Asante king, or *Asantehene*, Opoku Ware I (1700–1750) with the innovation. The king, he writes, ordered traders to "buy silk taffeta cloths in all colours. The artisans unravelled these so that, instead of red, blue, green, etc. cloth and taffeta, they had many thousand *alen* of woollen and silk thread." (The Danish alen was a unit about two feet long.) Whether or not the idea actually originated with the *Asantehene*, he certainly appreciated it. The multicolored cloth enjoyed royal favor and fed an aristocratic market. The prestigious new luxury cost ten times as much as plainer material.[45]

Although colorful, this cloth wasn't quite what we know today as kente. What distinguishes kente from other narrow-strip textiles are its alternating blocks of warp-faced and weft-faced weaving. These aren't the gingham-style checks you can make on any loom. Rather, vertically oriented patterns alternate with horizontally oriented ones. In the vertical blocks, only the warp threads show, completely covering the weft. In the horizontal sections you see only the weft threads, which conceal the warp.[46]

Creating the block pattern requires not only planning and skill but also special equipment: a loom with two separate pairs of heddles. You thread the front pair the normal way, with a single warp thread going through each heddle; one shaft lifts the odd-numbered threads, and the other, the even ones. This pair is for the warp-faced blocks.

For the back pair, by contrast, you put the threads through the heddles six (or sometimes four) at a time, alternating between odd and even bunches rather than individual strands. Packed together, the warp threads get covered by the weft. For the most luxurious kente cloth, known as *asasia* and reserved only for royalty, the loom adds a third pair of heddles, permitting twill diagonals.

Scholars continue to debate exactly how and where the "double-heddle" loom developed. Among ordinary Ghanaians, the origin of kente cloth is highly contested and informed by ethnic rivalries. Asante and Ewe both want to claim the national fabric.[47] In fact, kente most likely emerged from the fusion of weaving traditions.

In Asante territory, weavers supplying the *Asantehene* and his court clustered in the town of Bonwire, near the capital of Kumasi. Theirs was a hierarchical and tightly controlled industry, overseen by a chief known as the *Bonwirehene*. He maintained overall production standards and directly supervised royal weaving. He also ensured that no one breached social etiquette by buying cloth above their station. In the early 1970s, writes Venice Lamb, one of the first scholars to collect and document West African textiles,

> the Bonwirehene told me that fifty years or so ago he would have refused to sell a good silk cloth to a young boy or to a person of no social consequence, and it would have been considered a sign of disrespect towards the Elders for a young person to have been seen in public wearing such a cloth. Good cloths were for Chiefs and "big men" only.[48]

Serving this elite market, Asante weavers developed great skill in creating and executing silk patterns. Bonwire became a haven for ambitious artisans. But the royal patronage and centralized locale that fostered high-quality work also limited variation and innovative techniques.

Living on the savannah, the Ewe had greater access to cotton than the forest-based Asante. They also lacked their passion for brilliant colors. So, while they did incorporate silk, the Ewe wove even their finest cloth primarily from cotton dyed in subtle hues. Their weavers were also more spread out, independent, and market-oriented. The best Ewe products weren't restricted to royalty. Anyone with the resources could commission a custom-made cloth.

And there was a crucial difference in fabric structure. All Asante weaving was warp-faced. The Ewe, by contrast, produced both warp- and weft-faced textiles. In addition, their cloth often featured stylized figures—birds,

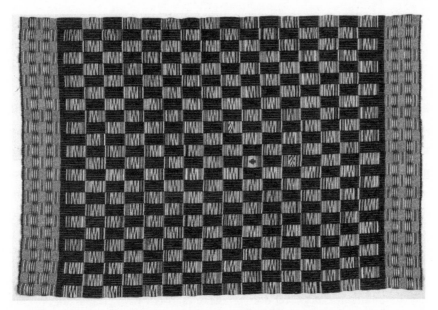

In a full kente cloth, individual strips alternating weft-faced and warp-faced blocks are woven and then sewn together. This Asante cloth was made of cotton and rayon or silk in the mid-twentieth century. (Courtesy of Indianapolis Museum of Art at Newfields)

fish, crocodiles, flowers, leaves, people—woven as supplementary weft. They had the experience, in short, to conceive of alternating warp- and weft-faced blocks.

Drawing on surviving textiles, missionary photographs, and linguistic analysis, textile scholar Malika Kraamer makes a convincing case that Ewe weavers were in fact the first to use two pairs of heddles to produce kente cloth's characteristic blocks. Once conceived, however, the innovation quickly spread. On the fringes of their home territories, Asante and Ewe weavers sometimes worked near one another, encouraging cross-pollination. And the Ewe's vigorous textile trade carried the new idea. Looking at the alternating blocks on an Ewe cloth, a clever Asante weaver might figure out how to construct them—or at least start asking the right questions.

Whatever the mechanism, by the mid-nineteenth century, both Ewe and Asante weavers had adopted alternating blocks, with each group putting its own stamp on the designs. The Ewe preferred sober hues and representational motifs. The Asante prized brightly colored cloths with geometric designs—eye-catching symbols of power and prestige.

It was these Asante designs, popularized by Kwame Nkrumah, the first president of Ghana, that became known internationally as kente cloth, the "uniform of pan-Africanism" and emblem of African diaspora pride. Nkrumah wore kente on his state visit to the United States in 1958, and *Life* magazine ran photos of him and his entourage in the distinctive attire during his meeting with President Eisenhower and other official ceremonies. When the American sociologist and civil rights leader W. E. B. Du Bois received an honorary doctorate from the University of Ghana five years later, he wore strips of kente sewn on his academic robes. Other African American luminaries, including Adam Clayton Powell Jr., Thurgood Marshall, and Maya Angelou, took up the idea. In 1993, the once-elite practice broke through to undergraduates, when West Chester University in Pennsylvania held a special "kente commencement ceremony" to honor its black graduates. Today, graduates at every level don kente stoles, often with class years or other lettering woven in.

"When Black students wear Kente stoles as a sign of their successful matriculation through higher education, they transform their bodies into living, breathing proverbs," writes historian James Padilioni Jr., a West Chester graduate now on the faculty at Swarthmore. Addressing graduates, he articulates the meaning now ascribed to the cloth:

> The Kente stole you don around your shoulders testifies to the ancient wisdom of Africa and the "dream and the hope of the slave." The Asante stylized their values and ethics through the poetics of Kente. Kente's Diasporic genealogy weaves a pattern of African knowledge and pride across the Middle Passage and onto the capped and gowned bodies of Black American graduates.

Beautiful, ingenious, symbolic, and distinctive, kente connects wearers in the diaspora to the motherland, real and imagined, and to their ideal selves. "When I wear kente cloth it says I'm from Africa. I'm royalty," says a New Jersey graduate.[49]

As kente cloth went global, it took on forms and uses its originators never imagined. "Kente has been one of the first fabrics to become synonymous with Afrocentric clothing," a Manhattan importer told the *New York Times* in 1992, when kente patterns were enjoying an upsurge of interest among African Americans. Beware of counterfeits, he warned consumers. "In making the authentic fabric," the *Times* explained, "cotton cloth is first bleached white and then the design is wet-printed with dye on both sides. Since that is a relatively expensive process, the fake prints are usually dyed on only one side." It didn't mention that, strictly speaking, authentic kente cloth isn't printed at all. It is woven from colored yarn and assembled into a pattern that takes far more planning than a cotton print.[50]

In response to tourist demand, kente weavers now produce strips never meant to be sewn together into larger cloth. Some serve as stoles or wall hangings, while others are cut up to make items such as hats or handbags. "If we can speak of a 'tradition' of kente, then these items must be included as part of that tradition," writes an art historian. "There are few, if any, textile 'traditions' on the planet that have such a dynamic history. Certainly there are few textile 'traditions' wherein a square centimeter of cloth covers the surface of a single bead to become an earring."[51]

Kente and kente-inspired earrings, bowties, and yoga pants might shock traditionalists, some of whom even disapprove of wall hangings and table runners. "Kente was woven to be worn," declares a Ghanaian scholar and social critic, decrying such home decoration as "culturally subversive." But trying to fix the form and function of textiles is both futile and unwise. Turning kente cloth into bedspreads is no more culturally subversive than adding foreign silk to please the king. Living textile traditions change, reflecting the identities and desires of the people who use the cloth.[52]

※※※※

It's early evening, and the woman in red looks like she's heading home after a day in the market of San Juan La Laguna, a town on the shores of Lake Atitlán in Guatemala. She is wearing a traditional ensemble, or *traje*—except for the smartphone tucked into her tightly cinched *faja*, a wide, handwoven sash. Drawn by the contrast between old and new, I ask a Guatemalan friend to ask if I can take her photo. Something gets lost in translation. Happy to cooperate, she removes the phone and hides it behind her back.

No, please tell her I want the phone in the picture. She proudly poses with it in her left hand. Still not part of the outfit. Oh well.

Although it includes the essential components that mark her as a Maya, the woman's *traje* isn't as traditional as it initially appears. Her top is not a handwoven cotton *huipil* but a factory-made blouse, probably polyester, adorned with machine embroidery and rhinestones—less expensive and more practical for everyday wear than the heavy cotton rectangles woven on a backstrap loom and sewn together. Her skirt, or *corte*, is the critical component of the outfit; the Guatemalan idiom "*Lleva corte*," or "she wears a [traditional] skirt," means a woman is indigenous. Yardage wrapped around the body and secured by the *faja*, hers looks like it came off a traditional floor loom—a technology introduced by the Spanish—but the red and navy plaid reflects fashion rather than custom. Her outfit is as up-to-date as her nail polish and cell phone. Yet it is still indisputably Maya.

In the well-established romantic narrative, material progress represents a devil's bargain: shoes, running water, and vaccines at the cost of beauty, identity, and meaning; uniqueness replaced by homogenized global culture. Maya *trajes* illustrate a different—and likely more common—pattern. Left to their own devices, consumers rarely treat tradition and modernity as all-or-nothing choices. They find ways to maintain their inherited identities, including the material manifestations that signify belonging, while satisfying the desire for novelty and self-expression.[53]

Contrary to the nostalgic vision of timeless peasant customs, Guatemalan textiles have always been dynamic. Many *huipiles* incorporate colorful designs in supplementary weft brocade, some geometric, others featuring stylized animals, plants, and people. The bright threads to make the designs initially came from Chinese silk floss—"There are five generations of Chinese in Guatemala," notes textile collector Raymond Senuk—and, when World War II interrupted the supply, weavers adopted shiny mercerized cotton.

Picked out row by row with fingers or a pointed stick similar to a knitting needle, the patterns range from ancient Maya imagery to contemporary innovations. At a shop in Antigua that sells secondhand *huipiles*, I bought one featuring rows of donkeys, rabbits, scorpions, roosters, quetzals (the national bird), baskets, spiders, humans, and, clinching the sale, helicopters! When magazines began printing cross-stitch patterns in the nineteenth century, Maya weavers adapted the designs, inventing a new form of brocade, known as *de marcador*, in which the supplementary threads wrap around the warp so that both sides of the fabric are identical.

The red that dominates many of the most traditional-seeming *trajes*—those worn for religious ceremonies—in fact dates only to the nineteenth-century introduction of alizarin (synthetic madder) dyes from Germany. Although madder grows in Guatemala, locals had never learned to use it, and they lacked the mordants needed to dye cotton with the region's famous cochineal.

Without giving up backstrap weaving, indigenous people also embraced European floor looms, using them to produce fabric for skirts, aprons, and trousers. Probably inspired by Asian fabrics, they developed a new tradition of dyeing called *jaspe*. Better known elsewhere as *ikat*, jaspe is a complicated tie-dye technique in which undyed threads are tied to block out a pattern that appears when the cloth is woven. (You can identify ikat by the slightly blurred appearance of the figures.) In addition, today's floor-loom cloth frequently incorporates metallic threads made of coated polyester film.

Detail from a Guatemalan huipil *whose supplementary-weft designs include helicopters along with more traditional symbols (Author's photo)*

Far from a dying art, says Senuk, "weaving is fine in Guatemala. But it's changing, really dramatically. In the last twenty years, dramatic things have happened." Until a few decades ago, you could easily identify a Maya woman's village simply by looking at her clothes. Although each weaver created her own patterns, they worked within well-defined rules of construction, background colors, and decorative designs. A *huipil* from San Juan La Laguna would feature twenty-four embroidered squares, arranged as four rows of six, below a yoke decorated with zigzags, all on a red striped background comprising two woven pieces. The accompanying *corte* would be black and white.

In the northern highlands village of Todos Santos Cuchumatán, by contrast, a *huipil* would be sewn from three panels, woven with alternating red and white stripes. The center section would have geometric patterns brocaded in supplementary weft, with a yoke adorned with store-bought rickrack. The stripes could be larger or smaller and the brocade designs could vary, sometimes spreading to the other panels. But to a knowledgeable observer the blouse would clearly declare that its wearer hailed from Todos Santos. Every village had its own distinctive combination of elements.

By the 1990s, things began to change, as women started to buy and sell clothes in local markets rather than making everything themselves. "I would see a woman in the market that I would know was from San Antonio Aguas Calientes and she would be wearing a *huipil* from the Alta Verapaz, from Cobán, and I would say, 'Porque?'" recalls Senuk. "And she would say, 'Because I like it.'" Picking and choosing from other villages' *trajes* evolved into new "pan-Maya" fashions that weren't specific to any particular place.

Around the turn of the century, Maya women invented the novel style that captured my attention on the street in San Juan La Laguna: the monochrome outfit, with *huipil*, sash, and skirt—and sometimes apron, hair band, and shoes—all in coordinated colors. "What you did was you now got a base color—say, turquoise," explains Senuk. "You now went and got a turquoise *huipil* that was machine embroidered with related colors. The skirt was an ikat skirt but with turquoise bands in it. And the belt was a Totonicapán-style woven belt but turquoise. Now you were either turquoise, pink, *café*, purple—and all these things were possible choices. And they had no village significance at all." Monochrome fashion makes it easy to produce a striking outfit that is simultaneously Maya and Instagrammable: #chicasdecorte.[54]

By the late 2000s, internet-savvy shoppers expected to find exactly what they wanted online. In a 2004 article and subsequent book, *Wired* editor Chris Anderson made "the long tail" shorthand for the phenomenon, writing that

> Our culture and economy are increasingly shifting away from a focus on a relatively small number of hits (mainstream products and markets) at the head of the demand curve, and moving toward a huge number of niches in the tail. In an era without the constraints of physical shelf space and other bottlenecks of distribution, narrowly targeted goods and services can be as economically attractive as mainstream fare.[55]

So when Stephen Fraser's wife couldn't find fabric with large-scale yellow polka dots for the curtains she wanted to make, her "internet geek" husband offered to search online. He came up empty; nobody sold what she had in mind. No problem, he figured. There must be a website printing textiles on demand, just like the self-publishing start-up where Fraser had been a marketing executive. No luck.

Soon, Fraser was having coffee with his former colleague Gart Davis, also married to an avid crafter, to talk about filling the empty niche. Before starting the company they dubbed Spoonflower in 2008, they went to the textile college at nearby North Carolina State University to check out its digital textile printer. The machine was reassuringly familiar. "I looked at it and said, 'That looks like an inkjet printer on my desk, just a little bigger,'" Davis recalls. "How hard could it be?"

It was ridiculously hard.

Textiles, it turns out, are much more finicky than paper. They're floppy and even what seems to be a standardized bolt of cloth encompasses subtle variations. "You almost have to caress the fabric. It's very artisanal," says Davis, showing me the company's original printers. "Getting five yards out of one of these beasts is tricky." In the early days, Spoonflower could print only two or three yards of cotton cloth an hour. But enough customers were eager to design and make their own fabric—and willing to pay a premium for the privilege—that the company survived.

Over time, digital textile printing improved, Facebook provided an ideal marketing vehicle, Spoonflower expanded its range of base textiles, and the company became a big deal in its little corner of the textile world. By late 2019, it had more than two hundred employees, divided between Durham, North Carolina, and Berlin, and shipped about five thousand pieces of cloth a day. Each averages about a yard.

"We're a tiny, tiny little company," Davis admits, "but on the internet nobody knows how big you are. They have this concept of Spoonflower as being this giant thundering Facebook-like thing in textiles."

For Jonna Hayden, Spoonflower is a godsend. "I'm a costume designer in a tiny town," she says, "with budgets that require fiscal discipline in the extreme." A decade ago, every show entailed frustrating compromises, since local textile sources were limited and Hayden couldn't dream of commissioning custom materials to fit her vision. Spoonflower now gives her choices once reserved for big shows in big cities. "I can *design exactly what I want* and upload it to a website, order a swatch for $5 and see it next week," she exults in a Facebook message.

> In effect, Spoonflower has removed the Guild status from the fabric designers and mill houses. I'm no longer subservient to what *they* decide are the trends or prints or colors of the year. I can make my own, to my personal desire, and achieve the look I want, not "close to" what I want.

Hayden is exactly the kind of customer Davis and Fraser had in mind when they started the company: someone with a vision who designs her own fabric for her own use. But she isn't their typical client.

Business really took off when Spoonflower discovered, more or less by accident, the good old economic phenomenon of specialization—and a much larger market. As a promotion, the site began sponsoring weekly design contests, challenging customers to create a print with cats, for instance, or something for Halloween. Customers voted for their favorites, the winners got credits they could use on the site, and the company sold small runs of the winning designs through Etsy's online marketplace. The Etsy results made it clear that the world was full of potential customers who weren't themselves designers—crafters, mostly, who wanted more choice than they could find in a typical fabric store.

Once again, textile consumers proved surprising. "I thought maybe 10 or 20 percent of our business would turn out to be marketplace business—a stranger meets a stranger," says Davis. It now tops 75 percent. With more than a million designs for sale, Spoonflower is serving the long tail.[56]

Using state-of-the-art technology, it's also reviving some of the preindustrial qualities of textiles. By replacing mass production with custom-made fabrics, Spoonflower enables consumers to visually define themselves more exactly. You can order patterns with Babylonian, Sumerian, or (yes) alien cuneiform, Nordic runes or Mongol calligraphy, the Hail Mary or the Shema. The chintz once forbidden in France comes in the blues and reds of Indian prints, neon pinks on black that would excite a Victorian seamstress, Pop Art repeats and photorealistic roses. You can buy toiles with traditional pastoral scenes or ones that sneak in images from *Star Trek*, *Doctor Who*, Agatha Christie, or the *Legend of Zelda*—or that honor women in science, suffragettes, or runaway slaves.

"What I'm hoping for," says Davis, "is that people connect with your tribe—whatever it is that you're trying to say, whether it's steampunk goth with a slightly red theme or it's Welsh language glyphs. You ought to be able to speak your tribe." Again and again, textile consumers remind us that cloth is more than just stuff. It is desire and identity, status and community, experience and memory embodied in visual, tactile form.

Chapter Seven

INNOVATORS

The important improvements and innovations in clothes for
the World of To-morrow will be in the fabrics themselves.

—Raymond Loewy, *Vogue*, February 1, 1939

WALLACE CAROTHERS DIDN'T SET OUT to create a new fiber, let alone an entirely new kind of material. He was trying to settle a scientific dispute.

A music lover and omnivorous reader, Carothers was above all a dedicated chemist driven to explore fundamental questions about the structures
of materials. In 1924, while still a graduate student, he published an auda-
cious article applying Niels Bohr's pathbreaking model of the atom to organic molecules. The paper was so controversial that the referees deadlocked
on whether to accept it. In time, it came to be regarded as a classic.[1]

Although devoted to pure science, with little talent for business or interest in engineering, in 1927 Carothers nonetheless found himself wooed by
industry. The DuPont chemical company was setting up a basic research lab,
and it wanted the thirty-one-year-old Harvard instructor to head its organic
chemistry section. Carothers was enthusiastic about the venture, but despite
the promise of a significantly higher salary, an able staff, and the freedom to

investigate whatever caught his fancy, he declined the offer. Academia, he said, better suited his unsettled temperament. "I suffer from neurotic spells of diminished capacity which might constitute a much more serious handicap there than here," he wrote the recruiters.

DuPont persisted, coming back a few months later with a higher salary. This time, Carothers took the job. The money, though welcome to the always-strapped young man, wasn't what changed his mind. In the interim, he had found a scientifically intriguing problem that he believed would complement his new employer's commercial interests: What exactly were polymers?

In answering the question, Carothers would do more than satisfy his chemical curiosity. He would set off the greatest materials revolution since the development of ceramics and metallurgy. His research exemplifies what economic historian Joel Mokyr, writing about an earlier period of technological progress, calls the Industrial Enlightenment. Pure science and practical craft tend to make the greatest advances—and are most likely to alter the textures of daily life—when they inform each other. The interaction gives basic researchers new tools to use and new questions to ask, while offering artisans, engineers, and entrepreneurs guidance on where to direct their attention. "Without DuPont's forcible approach and constant contact with him during late fall of 1927," observes a historian of science, "the young Harvard chemist might never have turned his attention to polymers, nor contemplated a new research program."[2]

Throughout history the desire for more and better cloth has driven technological innovation, from hybrid silkworms to digital knitting, from belt drives to bills of exchange. The sheer ubiquity of textiles—and the money to be made in producing and selling them—amplifies their influence. They spark the imagination of scientists and inventors, investors and entrepreneurs, mercenaries and idealists. Change textiles and you change the world.

𝄞𝄞𝄞𝄞

By the late 1920s, organic chemists understood that the building blocks of common natural substances such as proteins, cellulose, rubber, and starch—including all biological fibers—were much larger than the simple molecules with which they had established their science. Beyond that, polymers were a mystery. Most chemists believed these strange materials were not in fact single compounds but agglomerations of smaller molecules held together by some as-yet-unknown force.

Hermann Staudinger dissented. The German chemist argued that polymers were true macromolecules, thousands of times larger than the ones chemists were accustomed to working with. When he presented his theory at a 1926 meeting, the assembled organic chemists were stunned. "We are shocked like zoologists would be if they were told that somewhere in Africa an elephant was found who was 1500 feet long and 300 feet high," said one. It didn't help that Staudinger had little empirical evidence to back up his claims.[3]

Believing Staudinger was correct, Carothers set out to find the missing proof. The first step was to create macromolecules larger than any previously synthesized, using acids and alcohols to form the compounds known as esters. By repeating the reactions over and over again, the DuPont team built up long chains: the first polyesters, although not the specific compound we now know by that shorthand. The new molecules broke the size record, but the team still couldn't manage to top a molecular weight of 6,000—much smaller than those known for many biological substances. Maybe Staudinger was wrong after all.

Then Carothers had a brainstorm. Along with the polyesters, the reactions produced water. Maybe its components were bonding with parts of the polyester chains, breaking them apart to re-create acids and alcohols. They needed to get rid of every bit of the water.

Carothers acquired a finicky new piece of equipment known as a molecular still. Using it, his deputy Julian Hill managed to slowly extract the water, boiling it off in a vacuum and trapping it as it froze to a cold condenser. The process took days, but finally Hill had a tough, stretchy polymer that remained extremely viscous when melted, suggesting a high molecular weight. He touched it with a glass rod and got an unexpected result. "There was this *festoon* of fibers," he later recalled.

Carothers wasn't there for the eureka moment, when Hill and other researchers began to draw out filaments, winding down the lab's hallways as they tested—and celebrated—the new material. Lustrous, soft, pliable, and strong, the strands resembled silk. With a molecular weight of more than 12,000, the new substance comprised a long chain of esters held together by ordinary chemical bonds. Carothers and his team had validated Staudinger's theory.

In June 1931, Carothers published a definitive article titled simply "Polymerization," demonstrating that polymers were regular molecules of extraordinary, theoretically unlimited, length. He detailed techniques for synthesizing the macromolecules and laid out a vocabulary to describe their features. With this single publication, the thirty-five-year-old Carothers established the new field of polymer science. "After that article," said Carl Marvel, a prominent fellow researcher, "the mystery of polymer chemistry was pretty well cleared up and it was possible for less talented people to make good contributions to the field."[4]

At the American Chemical Society's annual conference that September, Carothers and Hill announced the world's first entirely synthetic fiber. The *New York Times* heralded the "synthetic silk" as a "new landmark in the progress of chemistry."[5]

Its immediate significance was scientific and inspirational, however, not commercial. The polyester melted at too low a temperature to be practical for textiles. Efforts to create a more durable polymer using amides instead of esters failed. Carothers turned to other topics.

But the deepening economic depression was beginning to curtail his freedom. DuPont needed a return on its research investment, and his boss thought fibers could be just the thing. "Wallace, if you could just get something with better properties, higher melting point, insolubility, and tensile strength, you could have a new type of fiber," he told his star scientist. "Look it over and see if you can't find something. After all you are dealing with polyamides, and wool is a polyamide."[6]

So, beginning in early 1934, Carothers abandoned his beloved pure research and set out to make a polyamide that would tolerate hot water and dry-cleaning fluid. After a few months of systematic trials, the lab had its first success: a silk-like filament that stood up to both. Further experiments found a way to synthesize it using benzene, a plentiful coal derivative,

*Nylon inventor Wallace Carothers
demonstrates his first big discovery,
neoprene, a synthetic rubber. (Hagley
Museum and Library)*

thereby making the new fiber affordable. By the end of 1935, the first nylon yarn was ready for testing.[7]

Three years later, it hit the market—not as a textile but in Dr. West's Miracle-Tuft toothbrush, advertised as "a toothbrush without bristles." With its clean, white, uniform, and nonporous artificial filaments, the new brush promised to "end animal bristle troubles forever." No more splitting, getting soggy, or breaking off in your mouth. Introducing the new miracle fiber to the public, DuPont executives described nylon as made from "coal, air, and water." One of the first major uses, they said, would be women's hosiery.

The big reveal came at the 1939 New York World's Fair, where DuPont's pavilion featured a model wearing nylon stockings. When the first four thousand pairs went on sale that October, they sold out immediately, and within two years nylons accounted for 30 percent of the market for women's hose. Having promoted nylon as more resistant to snags than silk, the company soon had to damp down public expectations that the new stockings would never run. Even miracle fibers have their limits.[8]

World War II temporarily diverted nylon away from consumer products to parachutes, glider tow ropes, tire cords, mosquito nets, and flak jackets.

When Allied paratroopers dropped from the skies to launch the Normandy invasion, they unfurled nylon parachutes. Someone, perhaps an astute DuPont PR man, called the new synthetic the "fiber that won the war."[9]

Nylon was only the beginning. The British chemist Rex Whinfield had long dreamed of discovering a synthetic fiber, repeatedly returning to the problem beginning in 1923. When Carothers published his results, Whinfield knew he'd found the key.

In 1940 he and his assistant James Dickson began their own ester syntheses. They used the "less familiar and long neglected" terephthalic acid as an ingredient, on the theory that its greater molecular symmetry would produce better results than Carothers's polyester experiments. Early the next year, they drew the first fibers from a "very discoloured polymer" that Whinfield dubbed Terylene. Its chemical name is polyethylene terephthalate, but nowadays we usually just call it *polyester*. It is the world's leading textile fiber, outselling even cotton.[10]

Carothers didn't live to see the success of nylon or the world-changing ripple effects of his work. On April 29, 1937, his lifelong depression finally proved unendurable. Early that morning, he checked into a hotel, took out the cyanide capsule he'd carried since graduate school, mixed the poison in a glass of lemon juice, and killed himself. He was forty-one years old.[11]

In a research career spanning a mere thirteen years, Carothers revolutionized organic chemistry and transformed the substance of daily life. His contemporaries immediately recognized the significance of that achievement.

Inspired by the soon-to-open New York World's Fair, in February 1939 *Vogue* asked nine industrial designers to imagine what the people of "a far To-morrow" might wear and why. You may have seen some of the magazine's mock-ups online: an evening dress with a see-through net top and strategically placed swirls of gold braid, for instance, or a baggy men's jumpsuit with a utility belt and halo antenna. Social media periodically rediscovers a British newsreel of models demonstrating the outfits while a campy narrator ("Oh, swish!") makes labored jokes. The silly getups are always good for self-satisfied smirks. What dopes those old-time prognosticators were![12]

The ridicule is unfair. Anticipating climate-controlled interiors, greater nudity, more athleticism, more travel, and simpler wardrobes, the designers

Shoppers throng to buy nylon stockings the first day they were on sale nationwide in the United States. (Hagley Museum and Library)

actually got a lot of trends right. Besides, the mockups don't reveal what actually made the predicted fashions futuristic. Looking only at the pictures, you can't detect the most prominent technological theme: new fabrics. Every single designer talked about textile advances. They knew that a new wave of breakthroughs had begun and anticipated more to come. Like the dyes before them, twentieth-century fibers would not be wrested from living nature but designed in labs. Once again, fortunes would be made. Once again, the textures of daily life would change.

For the next few decades, textiles again played a high-profile role in scientific and industrial progress. They were high-tech, inspiring space-age fashions and interior design. They liberated women from household drudgery. "Curtains that could be drip dried, uniforms that never needed ironing, and sweaters that could be washed without shrinking reduced domestic burdens," observes a business historian. When large numbers of American women entered the workforce in the 1970s, they came attired in easy-care polyester pantsuits. By the 1980s, however, attitudes shifted. Synthetic fibers were no longer novel and had become downright unfashionable. "Pity poor polyester," began a *Wall Street Journal* feature. "People pick on it."[13]

Over the succeeding decades, synthetic fabrics got better—softer, more breathable, less likely to snag and pill, more varied in look and feel. Today's state-of-the-art raincoat, dress shirt, or pair of tights would amaze someone transported from 1939, or even 1979, but nowadays we just expect it to work. The incremental innovations that reduce wrinkles, make hoodies breathable, or extend the life of upholstery cushions are invisible. Wicking T-shirts and stretchy yoga pants don't attract attention the way nylon stockings did. The very success of textile innovators obscures their achievements.

⚙⚙⚙⚙

Kyle Blakely and his roommate were hungry. The North Carolina State freshmen had skipped breakfast to attend an early-morning college fair, where they listened as engineering and business professors explained why students should choose those majors. The two friends were unpersuaded. Engineering was too intimidating—differential equations!—and business was too popular. They didn't want to get lost in the crowd. As nine a.m. rolled around, they were as undecided as they'd been when they woke up.

The university required them to hear pitches from three different colleges. With growling stomachs and no clear candidate for the third session, they did the logical thing. They set out to find free food.

"It was a big long hall, and we went poking our heads into all the rooms to find the college that had breakfast left over," says Blakely, now vice president for materials innovation at athletic wear maker Under Armour. The lonely folks at the College of Textiles still had ample supplies of doughnuts and orange juice. The roommates knew nothing about textiles, but they were happy to swap attention for sustenance.

Forty-five minutes later, they were sold. The textile curriculum included both engineering and business and offered a clear application for what students learned. The college was small and, Blakely recalls, "their job placement rate was out of control. It was like 98 percent." The impressive statistic reflects the quality of the school, widely regarded as the country's best textile program. But it also indicates a talent shortage. Textiles have an image problem.

In the unlikely event they consider it at all, ambitious young Americans assume the industry is a stagnant backwater where innovation stopped decades before they were born. They don't anticipate interesting work in vibrant locations, and the remaining US mills aren't encouraging. Many are in rural areas and, says Blakely, when you visit, "It is like a time warp. Everything is wood panels. That's scary. Who wants to go into that? Everybody wants to be in these Google-Apple environments."

Environments, that is, like the one where he works. Under Armour's Baltimore campus, a revamped former Procter & Gamble factory complex, features bright open spaces with enough industrial grit to convey authenticity and heritage. Employee amenities, including a state-of-the-art gym and foodie-friendly cafeteria, would be right at home in Silicon Valley—if tech companies were full of jocks. (Does Apple have a training room for boxers?)

The well-equipped labs include everything from computer-controlled 3-D knitting machines to "Torso Tom," a test dummy that looks more like an upright missile than a person and that emits jets of steam to simulate sweat. With life-size sports photos lining the halls, the place is a shrine to athletic achievement—and to innovation. "We have not yet made our defining product" declares a slogan on the wall.

Today's textile power users aren't the royal courts of old—or, for that matter, couture houses or fashionistas. They're elite athletes and outdoor adventurers, soldiers and first responders. Under Armour and rivals such as Nike relentlessly compete to find new ways to satisfy their lead customers' never-ending demands for heightened performance.

Under Armour got its start in 1996, when founder Kevin Plank, a football player at the University of Maryland, decided to make T-shirts from the same tight, silky microfiber fabric as his compression shorts. Unlike his sweat-soaked cotton tee, the shorts stayed dry during practice. The polyester fabric's fine filaments, no more than a tenth the diameter of a human hair, spread the water out, allowing it to evaporate quickly. Although initially skeptical of replacing their cotton workout shirts with "girls' spandex," other players quickly embraced the new material once they realized it did keep them drier. (That the tight shirts highlighted their sculpted physiques also proved a selling point.) "We didn't invent synthetics," says Plank, "but we did invent this application."[14]

Enabling his entrepreneurial insight were the decades of small improvements that created microfibers in the first place. It was "not one particular technology but many," explains a textile chemist. "Some involved changing fiber shape, some involved using chemical treatments to reduce fiber size, some involved new fiber extrusion technologies in which fibers have more than one polymer component."[15]

Quick-drying and soft, microfibers rescued polyester's battered image. The fabrics became so common that nowadays environmentalists worry about what happens when the tiny fibers break off and enter the water system when laundered. (To both measure and control the problem, Under Armour's test lab has developed a filter for its washing machines.) Finding equally pleasing alternatives, such as the bioengineered silk we saw in Chapter 1, is now a research frontier. *Sustainability* has become a watchword among textile scientists.

As he seeks to improve Under Armour materials, Blakely increasingly focuses on the earliest stages of the manufacturing process. Starting early provides more possible ways to add features. To develop a cooling fabric, for instance, the company worked with an Asian supplier to develop a yarn whose cross section maximized its surface area. It then infused the material with titanium dioxide, whose presence makes people exercising in hot, humid environments feel cooler.[16]

Under Armour is also redesigning yarn shapes in hopes of developing new water-repellent materials. To keep wearers dry, the industry has traditionally applied coatings, known as durable water repellents, or DWRs, to cloth or completed garments. But those coatings depend on chemicals called fluoropolymers that have come under attack as environmentally harmful.[17]

Eco-conscious consumers want alternatives. But they also don't want to get wet, nor do they want their jackets to be heavier, less flexible, or more expensive. Alternative chemical coatings simply don't work as well. So Under Armour is trying a physical barrier. "We want to develop yarns that interlock next to each other in cross section," explains Blakely, "to create a kind of an impasse where water can't get through mechanically. And then the non-fluoro finish just helps."

Finding such solutions may not be as high-profile as launching a new app, but it's definitely innovation—made all the more challenging by an intensely competitive business that doesn't enjoy Google-style profit margins. "What I want to see in textiles is for the industry to be recognized as progressive," says Blakely. "It's genuine."[18]

Beyond the advances that steadily improve everyday textiles, bolder experiments are in the works. In an era where hardware constantly shrinks, nanotechnologists manipulate individual atoms, bioengineering is both a scientific frontier and a model for thinking about new materials, and environmental concerns are a cultural imperative, the tiny fibers that make up so much of our world offer an enticing playing field for ambitious scientists. Much of their research will never move beyond the pages of esoteric journals. Some will find only niche applications. Others will inspire widespread adaptations. A few will reshape the fabric of our lives.

Like the designers polled by *Vogue* in 1939, we can only peer hazily into the future. But even a brief tour of current investigations, limited for practical reasons to the United States, offers hints as to what the future of textiles might look like. It also reveals a change in the relationship between pure science and industry practice. While earlier periods saw textile research spill over into other fields—as dye production inspired chemical advances and polymer fibers led to plastics and protein chemistry—now the exchange often goes the other way. Researchers start in other fields and, realizing the ubiquity and importance of cloth, begin to apply their work to textiles.

ᚱᚱᚱᚱ

Ranging in height from a few inches to more than a foot tall, the crystalline objects could be miniature spires in an Art Deco diorama of a futuristic city. As they rise from clear bases embedded with gray and black stripes, their square sides curve smoothly inward at the corners until the stripes, parallel at the bottom, taper together into a long, dark point.

The spires are striking souvenirs of a manufacturing process whose output appears utterly prosaic: a thin filament coiled around plastic spools. It

Fibers containing electronic components and the remains of the preforms from which they were made at Advanced Functional Fibers of America (AFFOA) (Greg Hren Photography for AFFOA)

looks like something you might find in a hardware or crafts store. Nothing special.

The fiber, Yoel Fink maintains, augurs a fabric revolution.

A materials science professor at MIT, Fink was also the founder of Advanced Functional Fibers of America, or AFFOA, a nonprofit consortium. Its 137 members include universities, federal defense and space agencies, and companies ranging from start-ups to multinationals. In a small industrial building on the MIT campus, AFFOA does its own experiments and testing. It also coordinates a network of members who've agreed to develop prototypes.

Unlike typical university researchers, AFFOA's affiliates are thinking systematically not only about the next interesting scientific question but also about the next step in turning their work into real textile products. The network provides access to product designers, spinning and weaving mills, and assembly plants that can translate materials science advances into prototypes—and in some cases invent their own applications.

By bridging research and industry, Fink set out to make textiles as capable and as frequently and predictably upgraded as laptops or cell phones.

He wants "to bring the rapidly evolving world of semiconductor devices into the slow evolving world of fibers." He isn't speaking metaphorically about industry cultures. He means getting chips, lithium batteries, and other electronic essentials into fibers. Not wrapped in them, mind you, or connected to them, but permanently and impermeably embedded in them.

The spooled filaments aren't as ordinary as they look. They contain what Fink, with typical twenty-first-century parochialism, calls "the three fundamental ingredients of modern technology, which are metals, insulators, and semiconductors." In our day, *technology* doesn't signify machines, chemicals, or most other forms of *techne*. It means software and chips.

The spires are in fact upside down. They began as rods known as preforms, each about two feet long. To create a filament, a preform goes into the top of a two-story apparatus called a draw tower, which includes a small furnace. As the rod passes through the furnace, it is drawn down into a thread as thin as a human hair, stretching the material to ten thousand times the original length. The crystalline spire is the nub remaining when the fiber is clipped off from the bottom of the preform. It's a mere leftover—pretty trash. But it's big enough for visitors to look at and handle. The real action is too small for the naked eye.

Preforms and draw towers are well-established technologies used to make optical fibers like those that carry most internet data. Fink first worked with them two decades ago, when he was a newly minted PhD seeking to create hollow-core fibers lined with special mirrors to carry laser light. That invention, based on his doctoral research, led to a company called OmniGuide, which makes laser scalpels that cut and cauterize tissue with precision measured in microns.[19]

Bob D'Amelio, AFFOA's director of manufacturing operations, spent nearly thirteen years working at OmniGuide. Less than a year after departing for another company, he was lured back into the Fink fold. An enthusiastic East Boston native with a distinctive *R*-dropping accent and red Harley-Davidson T-shirt, D'Amelio walks me through the process that makes AFFOA's fibers special, using one carrying LEDs as an example.

Each preform is itself a precision-manufactured object. First, you cut a bar a couple of feet long and an inch or two wide from a thermoplastic, in this case polycarbonate. (The specific substance varies with the project; what's essential is that the viscosity changes continuously with temperature.) You then mill two narrow channels along its length, insert a placeholder known as a mandrel wire in each slot, lay a thinner sheet of polycarbonate on top, and fuse the top and bottom in a hot press. The result is a solid bar with mandrel wires hanging off the end.

Next, you drill a line of hundreds of individual pockets, each so small it's barely visible, into the thin sheet fused on the top of the bar. Using tweezers and a microscope, a technician drops a microchip into each tiny pocket. (A machine could do this job, but AFFOA doesn't yet produce enough to justify the expense.) Once the chips are in, you place another thin sheet on top. Finally, you take a second milled bar with a single channel and mandrel wire inserted and stack it on top. You hot-press the whole thing together, then remove the placeholders and feed permanent wires into the slots. The preform is now ready for the draw tower.

As the material stretches into a filament, says D'Amelio, "It's like, Honey, I shrunk the preform." Each component stays where it was relative to the others while the surrounding material gets smaller. Eventually the fiber is so narrow that the wires touch down onto the chips, creating a connection to carry electric power.

Getting the placement right demands the utmost precision. "Those wires have to touch down on the pads of the chip perfectly," says D'Amelio. "We're literally making a solder connection inside that oven." The materials and temperatures also have to be just right so that the thermoplastic melts sufficiently while the electronic components aren't damaged.

Variations on the same procedure can embed filaments with lithium batteries, sensors, or microphones. They can create fibers whose hues, like the colors of feathers, come not from dyes or pigments but from the way they filter light. D'Amelio, repeatedly marveling at his own workplace, calls the process "science fiction stuff."

The goal is to make the technology disappear into ordinary textiles, giving everyday cloth the ability to sense, communicate, measure, record, and respond. "We don't use the term *wearables*," says Fink, when I carelessly let it slip. "It's the word invented for stuff you don't wear. The stuff you wear you call clothes." He wants to turn textiles into platforms that spark the imaginations of countless inventors, just as smartphones led to unexpected applications.

AFFOA's fibers wouldn't replace regular yarns but would work alongside them, adding new powers to knitted or woven fabrics. They might let you stick your phone in your pocket to charge it from an invisible fiber battery or take calls on microscopic speakers and microphones embedded in your jacket collar. Your hat could give you directions while your underwear continuously monitored your health. Among the prototypes on display is a pair of tailored trousers that flash LEDs when they're hit by light: discreet safety apparel for walking after dark.

Materials scientist that he is, Fink believes the key to making textiles more powerful is changing their composition. "Fibers have always been made of a single material, and nothing that we know of that is advanced or can change rapidly year over year is made up of a single material," he says. "The number of degrees of freedom is just too small." Most fibers took millennia to reach their current state, and synthetics took decades. Fink wants to speed things up.

He envisions what he calls a Moore's Law for Fibers, named for the rule of thumb that computing power increases exponentially, with the number of components on a chip doubling every eighteen months to two years. Moore's Law isn't a law of nature but a self-fulfilling prophecy, driving innovative efforts and customer expectations. Each generation of chips is far more powerful than the previous one but less expensive per bit. So the price of computing power keeps plummeting, and people who write software and make electronic products plan accordingly.

The original Moore's Law itself enables AFFOA's fibers, by making their components tiny and cheap. What used to be expensive custom products from highly specialized manufacturers are now off-the-shelf commodities.

With semiconductors as a model, AFFOA's team pushes to hit their own progressively higher benchmarks. When I visit the lab, they've recently achieved one of those goals: getting fibers to survive more than fifty wash cycles, admittedly with cold water and no detergent. ("I have a feeling they'll be Dry Clean Only," remarked then chief product officer Tosha Hays, an apparel industry veteran whose father and grandfather owned a Georgia cotton gin.) Other indicators track such critical characteristics as bendability and tensile strength.

In her three years before leaving AFFOA in December 2019, Hays saw significant progress. "When we started we couldn't manufacture anything. Now we manufacture thousands of meters a day," she says. Prototypes "used to look like science projects. Now you have to look really hard to see the switch or the battery" on a fabric sample. Fiber diameters have shrunk by two-thirds, from one millimeter to three hundred microns.

But the filaments are still rigid and hard to work with. Wrapped in recycled polyester, they may look like regular yarn, but they are nowhere as flexible and resilient. Creating prototypes requires clever thinking about fabric structures and garment construction. You can't simply weave or knit them in.

Commercial applications will need the smart fibers to be much easier to work with. Making them thinner will help but, as Fink admits, "Without stretch we're not going to get very far." One of his graduate students is attacking the problem, he assures me, and "a year into it, she's pretty much got it figured out."

Then there's the challenge facing all electronic devices: power. Smart devices get dumb fast without electricity, and AFFOA's charging drawer is disconcertingly crowded with batteries sucking up juice. The team is working hard on energy storage. They can already make fibers containing lithium-ion batteries or super-capacitors. Super-capacitors don't hold as much power as batteries, but they charge much more quickly, don't wear out, and—an important consideration for textiles—don't spontaneously catch fire.

Building batteries into fabric yarn would eliminate the weight they now add, a major plus for soldiers in the field and a benefit for athletic wear and medical applications. Hays points to a mannequin wearing running tights studded with tiny lights. "Imagine you have those tights but there's not a separate battery," she says. "It would be in the fabric itself." The challenge is to make the battery fibers flexible enough for real-world clothing and dependable enough to rely on for essential power.

Even fiber batteries still need to be charged, however. You only have one smartphone to keep powered. You have a whole wardrobe of garments. "If you look at the term 'smart textiles,'" says a skeptical apparel executive, "how smart is a textile if you have to plug it into something?"

So maybe AFFOA's high-tech fibers will only go into specialized products and never make it to everyday apparel. Maybe fiber batteries will charge as you walk around, using the energy generated by your motion. Or, suggests Fink in a sci-fi moment, "there could be an inductive coil in your chair and an inductive coil in your pants. That's actually not so far-fetched. You are in contact with fabrics all the time."[20]

▨▨▨▨

Fink and Juan Hinestroza are both materials scientists. Both graduated from college in 1995. Both came to the United States as PhD students, Fink from Israel, Hinestroza from Colombia.[21] Both run labs at prestigious universities: Fink at MIT, Hinestroza at Cornell. Both do research that could transform everyday textiles.

But they are in other ways complete opposites. Their differences illustrate the ferment, and uncertainty, surrounding the future of textiles.

Fink is as much entrepreneur as scientist, sharply focused on turning research into prototypes with commercial potential. He advocates "Shot Clock Innovation," with ninety-day research grants and progress measured in weeks. As he talks about "Moore's Law for Fibers" or states that "fabrics occupy potentially the most valuable real estate in the world—the surface of our bodies," he demonstrates the self-assured sound bites that would go over well in a TED Talk. You can imagine him successfully pitching elected officials and textile executives to get AFFOA started.

Hinestroza isn't watching a shot clock. His research takes years, and he clearly distinguishes his role from that of industry. "We do pre-competitive research," he says. "Academic questions are not easy to solve, and that's why they're in academia." He explores fundamental science, patents relevant results, and leaves development to outsiders. Although much of his work is inspired by practical problems, he doesn't try to apply it himself.

While Fink converts chunky bars into fine threads, Hinestroza operates at a much smaller scale. He manipulates molecules to create new coatings, or *finishes*, as they are called in the fabric trade. His work unites textiles and nanotechnology—"large areas and microscopic elements," he says, "visible and invisible." Rather than turn fabrics into platforms for meeting as-yet-undefined needs, he seeks to make them better at performing their traditional functions of protection and adornment. He starts with well-established problems and looks for new ways to address them.

Fink believes that getting twenty-first-century technology into textiles demands new fibers. Hinestroza disagrees. People like the way existing materials look and feel. Established fibers, he observes, are also better able to survive the "extremely violent" stresses of looms and knitting machines. "I think that we can solve those issues with the same fibers that we have been using for thousands of years," he says. He works with cotton.

Hinestroza conducted his first textile research as a rookie professor at North Carolina State in the early 2000s, amid post-9/11 fears of chemical and biological weapons. Protective clothing then relied on different layers for different chemicals, an inherently limited approach. "You cannot have more than five T-shirts," he explains, using a favorite garment as an example. "I decided to make these T-shirts one molecule thick. By making them one molecule thick, I could make thousands of them and address bigger threats of different chemicals." Instead of piling up different textiles, he would layer protective chemicals on top of a single fabric. One molecular layer could be designed to trap mustard gas, another to block nerve agents, another to block bacteria, and so on.

To build up the protective molecules, he adapted techniques originally developed for making semiconductors. But unlike uniform silicon wafers, cotton is highly inconsistent. The cellulose polymers may all be the same, but characteristics like the degree of twist in each fiber differ. "The cotton in Alabama is different from Texas, is different from Vietnam," he explains. "The one in one year is different from the other. The one from the same year is different depending on the fertilizer that you use. We had to overcome all these issues."

Two decades later, Hinestroza has moved to upstate New York. But he is still fascinated by cotton. It is harder to work with than engineered polymers such as polyester and nylon, making it more scientifically challenging. And its deep history appeals to him. "We have developed a unique relationship with this fiber," he says. When he asks audiences to raise their hands if they *aren't* wearing cotton next to their skin almost no one does—and those who do are usually mistaken.

Hinestroza is now working on what he calls a "universal finish" for cotton. To give fabrics desirable qualities, such as stain resistance or wrinkle reduction, textile manufacturers typically apply chemicals to the cloth after it's made. The "permanent press" clothing that excited ironing-weary housewives in the 1960s came from new finishes, and such coatings still account for many incremental textile improvements. When men's no-iron khakis hit stores in the 1990s, just in time for business casual dressing to take off, a new finish was the crucial innovation.

Wander through the aisles at the giant Outdoor Retailer show and you'll find apparel brands touting all kinds of functional textiles, including clothes that stand up to abrasion (think rock climbing), re-

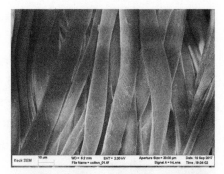

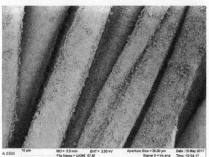

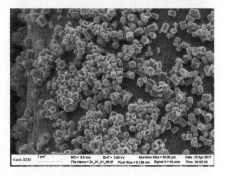

Bare cotton and cotton fibers with metal-organic frameworks attached, at increasing levels of magnification (Cornell University, Juan Hinestroza)

sist water, or repel insects. There are quick-dry shirts and socks that pledge not to stink. Many of these effects depend on chemical finishes. So do wools that don't shrink, hospital upholstery that kills germs, and sweaters that promise fewer pills.

But finishing is an inexact process that requires different treatments for different effects and that, in some cases, such as fluoropolymer water repellents, raises environmental concerns. Finishes also don't last. Over time, they wash out and rub off.

Hinestroza envisions a single all-purpose molecule that stays permanently bonded to cellulose polymers—the structural backbone not only of cotton but also of ancient plant fibers, including linen and hemp, and of rayon (viscose) and its more ecofriendly cousins modal and Tencel. He is using what's known as *reticular chemistry* to create an invisible net attached to every cellulose molecule.

"Since we understand each component of the cellulose molecule or of the polyester or nylon molecules, we can design molecules to exactly 'lock' into these features," he explains. Like puzzle pieces, I suggest. "A puzzle is a good analogy, only the pieces of the puzzle are coated with 'crazy glue,'"—they chemically bond—"so once in place they never detach from it."

Making up the net are molecules called *metal-organic frameworks*. As the name suggests, each has two components: organic molecules and metals. Picture a hexagon with a metal ball at each corner, connected by chains along the edges. The chains are organic compounds. The interior constitutes a "cage" that can hold other substances. You can make the space larger or smaller by altering the length of each chain, but the overall shape stays the same. Metal-organic frameworks are precise, predictable, and uniform. If they're hexagonal, they stay hexagonal. If square, they stay square.[22]

"That is the magic behind reticular chemistry," says Hinestroza. "One can reproduce in high fidelity a network of frameworks," which is ideal for making a consistent coating over thousands of yards of fabric. It's also possible to predict a framework's behavior mathematically, before actually synthesizing it—a major time-saver for difficult problems. "That's my favorite molecule," he says.

To create a cotton finish, the first step is to design a molecule that will bond to the cellulose polymer. Then comes the tricky part: finding a "package" or "load" with the desired functional properties that will fit in the cage and release when needed to do a particular job. The trigger could be abrasion, temperature or humidity change, or exposure to oils or bacteria. Releasing tiny amounts of finish in exact locations means the effects of the finish last longer.

It's a tough problem, with many variables. An oil-resistant coating must interact with many different kinds of oil; the same goes for antibacterial finishes. A water-resistant finish needs to let sweat escape while keeping rain out. Hinestroza deploys a graduate student to attack each puzzle.

So far, the team has managed to create a finish that simultaneously resists oil, water, and bacteria. But wrinkle prevention, a more complicated challenge requiring higher temperatures, remains elusive. Hinestroza would also like the treatment to repel mosquitoes, kill bedbugs, and, for certain applications, provide vitamins or medicine on demand. It might even replace dyes and pigments. When he says "universal finish," he isn't kidding.

Raising the degree of difficulty, he also wants to apply it using the same finish baths textile manufacturers already use. The goal is to give industry a precise method for producing more functional textiles with much less waste and no new capital investments. Finding a universal finish requires more time than a textile or chemical company could afford. "That's the beauty of academia," he says. "We engage in problems that are very difficult and that have not a trivial solution. If something has a trivial solution people will have found it before."

If he succeeds, one result would be so radical he doesn't really expect people to accept it. A textile that kills bacteria and repels oil wouldn't need cleaning beyond brushing off surface dirt. You might have to occasionally refill the "package," but there would be no more laundry. "If it doesn't stain, if it doesn't absorb outside compounds, then you don't need to wash it," he says. "But psychologically you will not be willing to wear something like that."[23]

᠅᠅᠅᠅

Greg Altman and Rebecca "Beck" Lacouture didn't start with textiles. But they did start with silk.

The two friends met on Lacouture's first day at Tufts University, where Altman was the graduate teaching assistant for the introductory bioengineering class. Fascinated by his research developing silk substrates to replace damaged knee ligaments, she served as a lab assistant and, after getting her own PhD, joined his start-up company, Serica Technologies Inc.

Serica adapted Altman's doctoral research to create mesh scaffolds for supporting soft tissue during and after surgery, particularly breast reconstruction. The scaffolds are made of silk that has been purified to remove the sticky (and sometimes allergenic) *sericin* protein that holds cocoons together, leaving only the *fibroin* that gives the fiber its distinctive strength and shine. Because fibroin is a protein, it is biocompatible and breaks down into amino acids that the body can easily absorb once the damaged tissue heals.

In 2010, Serica was bought by Allergan, the big medical company best known for Botox.[24] Altman and Lacouture stayed on for a couple more years before starting a new venture, also based on silk chemistry. This time, they wanted to make products that could have a broader impact than a niche medical device. Skin care, they thought, would be perfect.

When she was twenty-seven and three years out of graduate school, Lacouture was diagnosed with ovarian cancer. She recovered, but the chemotherapy was rough. It heightened her concern about how everyday substances might hurt someone with a compromised immune system. "Clean out your cosmetics cabinet," her oncologist advised at the beginning of the aggressive treatment.[25] Inspired by that experience, she thought silk—or, to be precise, fibroin—would make a good alternative to synthetic skin care ingredients. The protein is flexible, its crystalline structure catches light, it's smooth without being oily, and it's biologically benign. You can eat it without harm.[26]

Before they could put fibroin into moisturizers and skin serums, Altman and Lacouture had to persuade it to stay dissolved in water. Academic researchers had managed the trick for short periods, but the stuff inevitably congealed into a gel. The same thing happened here. "Our first passes were actually making silk Jell-O and then grinding it up," says Lacouture.

They attacked the problem with what Altman, who captained the Tufts football team, calls "simple blocking and tackling: Control the process. Measure what you have. Control the process. Measure what you have." The experiments took a year, but they eventually found the answer. The slightest impurity, it turned out, would give the fibroin something to bind to, leading it to fall out of solution. "Even salts from our town water supply," says

Altman, "can change how the silk might organize itself." Keeping the water absolutely pure was the secret.

After two years of work, their first skin care products hit the market. By that time, however, Altman and Lacouture had realized they shouldn't be in the skin care business—or at least not *only* in the skin care business.

"We realized that the chemistry platform that they created in the context of skin care is actually the thing of the greatest value," says chief financial officer Scott Packard, as he gives me a tour of the labs where thousands of waste cocoons are refined into what the company has trademarked as "Activated Silk." Skin care was just one possible use. "We could use that chemistry platform across a variety of textiles with a variety of applications. That's where it went from cool to a big deal."

Fibroin has an unusual chemical property. It contains some protein sequences that love water and some that hate it. The hydrophilic sections of fibroin allow it to dissolve in water, at least temporarily, and make it elastic. The hydrophobic sections bond to each other, thereby creating the fiber's strength. "There are sections of the silk protein that are so averse to being in water that they will stick, nearly permanently, to each other," says Altman. But they don't have to. They can stick to other substances instead. Hence the Jell-O problem created by impurities—and the aha business moment.

If fibroin would bind to salts in the tap water, it would bind to nylon, to wool, to cashmere, to leather—to anything that needed protection, softness, or other surface improvements. By cutting the fibroin polymers into different-sized pieces, some with hydrophobic edges and others with hydrophilic ones, the company could create different structures and material characteristics. It could fill in the rough spots on cheap leather, make nylon water-repellent and wicking, or reduce the tendency of wool and cashmere to shrink and pill. Just as it replaced synthetic skin care ingredients, silk could provide an alternative to existing textile finishes.

Skin care products touch some people. Textiles touch everyone.

Not that the company expects to capture the entire market. Now called Evolved by Nature, it is betting on a cultural trend: consumer revulsion against industrial chemicals. It aims its silk-based finishes at textile and leather producers looking for alternatives. "In a business sense," Altman acknowledges, "where there is no mission for sustainability, there is no ethics or morality, we're offering them nothing. Because everything we can do you can find a synthetic chemical to do."

As Altman's pointed language suggests, the company is on a crusade, one that attracts positive attention as the fashion industry excoriates itself for contributing to pollution and throwaway consumption. In June 2019, Chanel burnished its environmental image by buying a minority stake in the company.[27] The founders are, however, pragmatic businesspeople who understand that demanding a revolution won't move product or change industry practices. Nor does Evolved by Nature want to limit itself to luxury consumers. "We don't want to be like Tesla. We want to be the Toyota hybrid," says Altman. "I like the Prius and that's what we want to be."

To maximize their market, Evolved by Nature's finishes don't require new equipment or procedures. Workers simply follow instructions about how much to use and how much to dilute the solution in water, just as they do with other finishes. Although the prospect wouldn't thrill the company's staff, mills can even use Evolved by Nature products along with synthetics. "We're not disrupting the industry," says Packard. "We're trying to clean up the industry, one step at a time."

▨▨▨▨

In her white ensemble, she looks every bit the ageless fashion icon. Her straight skirt, falling to mid-calf, ends in a precise fringe, its sharp lines contrasting with her softly knitted bustier. A scarf tossed insouciantly around her neck completes the look. Its navy borders and the matching ribbon in her upswept hair add a touch of sophisticated color.

You won't find this Barbie in the toy department, however much you crave a respite from pink. She and her one-of-a-kind outfit live in an office at MIT, close to the university's famous dome. Research scientist Svetlana Boriskina uses Barbie to model her vision of future apparel.

An expert in optical materials, until a few years ago Boriskina would have dismissed clothing as "an old technology" unworthy of investigation. She focused on potential ingredients for cutting-edge devices.

Barbie's polyethylene ensemble keeps her as cool as being naked. (Svetlana V. Boriskina, MIT [psboriskina.mit.edu])

Then she received a challenge from the US Department of Energy, which was soliciting new ideas for "personal thermal management." It's a distinctly modern concept. For most of human history, after all, keeping warm was mostly a personal business. It meant wrapping up in clothing or blankets, with fires providing secondary heat. Air conditioning was unknown. Now, we rely on centralized climate control systems. They've improved our comfort, but heating and cooling buildings consumes a lot of energy—12 percent of that used in the United States. Technologies that let each of us live at our preferred temperature would reduce carbon emissions, with the side benefit of ending the endless conflicts over thermostat settings.[28]

Boriskina was intrigued. Her optical expertise suggested a novel approach.

Your body is constantly throwing off heat. Clothes trap it near your skin. Like bird feathers or animal fur, garments block *conduction*, which is the energy transfer that occurs when surfaces with two different temperatures meet. The excited molecules of the warmer material transfer energy to the more stationary molecules of the cooler substance until the two reach an equilibrium. Leave a frosty glass of beer on the bar too long and it warms to room temperature. Pull off your gloves to send a text and your fingers grow cold in the winter air. Some substances are especially rapid conductors, which is why jumping into a swimming pool on a hot day instantly makes you feel cold—and why you get quickly used to water that first felt chilly.

By putting a barrier between your warm skin and the cooler air, clothing blocks conduction. So when it's cold, the right clothes can already handle your thermal management, as Danish bicycle commuters are quick to point out. The real challenge—and what captured Boriskina's imagination—is cooling.

The solution needed to be passive, she thought, with no wires or batteries. Something simple. Something like clothes. Or like going naked.

Strip off your clothes and conduction will cool you down, as long as the surrounding air isn't hotter than your body temperature. (Sorry, Las Vegas.) But a second process works regardless of air temperature: *radiation*. Like the sun, your body is constantly pouring energy out into the universe. Whereas some of the sun's energy radiates as visible light, your body's doesn't; it has longer, infrared wavelengths, the ones night-vision scopes and infrared cameras pick up to spot people hiding in the dark.

About half the heat that leaves your body does so through radiation. But your clothes block it. "They just gobble it up," says Boriskina, "and then they warm up themselves and they trap heat around the body." On a hot day, clothes make you hotter.

But what if they didn't? What if textiles could be made of something that was transparent to infrared radiation—so that heat passed right through it—but looked opaque to the eye? Your clothes could let you feel as cool as you'd be without them while protecting you from sunburn and stares. Boriskina set out to discover whether such a substance might exist.

"We started purely from a concept," she explains. "It was mathematical: If we have this material, if we can cast it into fibers of a certain shape, and we put those fibers into yarn, it has the functionality we predict. And if you have that functionality, you can control the temperature."

The answer turned out to be a very simple polymer, just carbons and hydrogens, with none of the ionic bonds that tend to vibrate and block radiation. So far, so good. If you actually want something people will wear, however, you need more than the right equations. "It has to be comfortable," says Boriskina. "It has to be cheap. It has to be lightweight."

The right material not only exists, she discovered, but it's incredibly common. It makes up machine parts and pipe fittings, playground slides and shampoo bottles, recycling bins and the much-maligned disposable grocery bag. It has been called the "world's most important plastic," accounting for almost 30 percent of all plastic production.[29] It is polyethylene.

The one thing polyethylene isn't used for is making textiles. In fact, Boriskina had a hard time finding fiber for experiments. She finally located a Tennessee company called MiniFibers Inc., which chops polyethylene into bits for melting between surfaces as a kind of glue. They gave her enough uncut filaments to get started. You can easily extrude polyethylene on the same machines used for polyester, however, and Boriskina now has yarn made at the nearby US Army Natick Soldier Systems Center, which supports her research. She then uses AFFOA facilities to turn it into fabric. That's how Barbie got her clothes.

Lightweight, soft, and almost as cool as being naked, polyethylene sounds like a great textile, especially for hot, sticky climates. Why isn't it in our clothes? When I ask, textile industry veterans tell me it's too expensive, it degrades at high temperatures, and it can't be dyed.

As a general statement, the first objection isn't accurate. It refers specifically to an extremely strong, high-density polyethylene used for specialty applications. (Think heavy-duty marine chains that float on water.) Most polyethylene is incredibly cheap. That's why it's used for throwaway packaging.

The second criticism is true but might not matter, at least for some applications. Low-density polyethylene melts at around 130 degrees Celsius (266 degrees Fahrenheit), compared to polyester's melting point of around 260 degrees Celsius (500 degrees Fahrenheit). In water near the boiling point, it could shrink or otherwise misbehave. All things being equal, polyester is better. But in a hot climate, all things aren't equal. You could wash polyethylene clothes in warm water, hang them on the line, or even stick them in the dryer on a low-heat setting. You just wouldn't want to boil or iron them.

The dye problem, however, is absolutely true. It's the reason Barbie's outfit is mostly white. The polyethylene molecule offers no place for dye to bond, so the color just sits on the surface. When Boriskina's team tried coloring their miracle fabric with normal dye, it turned black. "Then we put it in cold water—whee!—and it's white again," she says. To get colored polyethylene textiles, you have to introduce colorants when you're making the fiber. That means less dye pollution—a big environmental plus—but it also changes the economics of the process, forcing mills to accurately forecast color demand. A fabric that can't be boiled and that is hard to dye isn't likely to take over the world. But a *cooling* fabric that can't be boiled and that is hard to dye still has some big potential markets, from abayas to underwear. It might even bring back the white workout T-shirt.

And the flip side to the dye problem is that stain molecules won't stick either. Cleaning is quick and easy, with no need for long wash cycles or high temperatures. Aside from the energy savings, you can imagine the benefits for people living without washing machines, especially in hot climates. The same antibacterial properties that famously keep polyethylene from degrading in landfills mean it's less likely to stink as a clothing material and could reduce infections in medical settings. It's easy to recycle, and the channels are already established. Old shirts could become new bottles, and vice versa. "The only thing you need to do is not throw it in the ocean," says Boriskina.

Boriskina didn't get the research money she vied for from the Energy Department, but its challenge changed her life. She has become a polyethylene evangelist.

"If you look now at fabrics—wearable or bed linens or tablecloths or car seats—they're all made now of either cotton or polyester. These two dominate almost the whole market. This is by any measure better than either cotton or polyester," she rhapsodizes, "and it's not more expensive to make. I don't see why it wouldn't replace almost any material." She is now researching its antibacterial properties and, with AFFOA funding, investigating whether polyethylene fibers could carry sensors or other electronics.

Her enthusiasm stems in part from the conviction that this material, which is often maligned by environmentalists, could do wonders for the planet. Turning polyethylene into an everyday textile, she believes, could save energy expended on air conditioning and laundry, even as it improved the comfort and health of people around the world. "That's what got me so excited in the first place, because it was totally not my field," she says. "You can actually make a difference in the world."[30] Working with textiles offers a chance to make a big impact. They may go largely unnoticed, but they are everywhere.

Afterword

WHY TEXTILES?

Old and new make the warp and woof of every moment.

—Ralph Waldo Emerson, "Quotation
and Originality," 1859

PEOPLE WHO HEAR ABOUT THIS book always ask me the same question: Why textiles?

I could give the kind of answer I imagine questioners expect. I could say that I grew up in a town that styled itself the "Textile Center of the World," so textiles loomed large in my formative years. But that's not really true. I had friends whose parents worked in the industry, but my personal contact was limited to an occasional trip to a factory outlet in search of cheap clothes.

I could say that someone in my family has been in the textile business for at least five generations. That, too, would be misleading. My engineer father moved from synthetic fibers to polyester film when I was still too young for kindergarten. His uncle in the carpet business died when I was little. Neither sparked an interest in textiles.

Much as it would make a good story, this book wasn't inspired by family history or childhood experience. It arose not from the familiar but from the strange: from calico prohibition and brazilwood exports, Minoan tablets and Italian throwing mills, a nineteenth-century dress with brilliant purple stripes and the promise of silk grown from yeast. My textile exploration originated in wonder. As I heard from scholars, scientists, and businesspeople—at first coincidentally and later as I began researching the subject—I was repeatedly struck by what a fundamental technology textiles represent, what world-shaking consequences they've had, and how remarkable much of their history is.

Exploring textiles introduced me to amazing natural phenomena like the weird chemistry of indigo and the improbable genetics of cotton. It showed me the ingenuity and care behind both handcrafts and industry—the pattern strings of a Lao loom and the nylon they're made of, the multiple stages of Indian block printing, and the thousands of yards flowing through a Los Angeles dye house. It made me grateful for the Industrial Revolution's bounty of thread and the women's time it freed.

I admired the enterprise with which Italian merchants created a mail service and their African counterparts turned strip cloth into currency. I laughed at the Tlaxcala councilors fretting about cochineal *nouveaux riches* and pictured the young Machiavelli doing word problems about bolts of cloth. I cheered Agostino Bassi as he tenaciously sought the origins of silkworm disease and grieved at the loss of Wallace Carothers. I felt the frustrations of Lamassī and smelled the stench of murex dyeing.

I shuddered at the ruthlessness with which Mongols marched captive weavers across Asia and Americans did the same to enslaved workers bound for the Mississippi valley. I wondered what would have happened if the provisions of the Northwest Ordinance, which included a ban on slavery, had been applied to all new additions to the original thirteen states. With different choices, could cotton have meant opportunity and liberation?

The more I learned about textiles, the more I came to understand about science and economics, history and culture—about the phenomenon we call *civilization*. We suffer textile amnesia because we enjoy textile abundance.

And that amnesia exacts a price, obscuring essential components of the human heritage, hiding much of how we got here and who we are.

Every scrap of cloth, I now realize, represents the solution to innumerable difficult problems. Many are technical or scientific: How do you breed sheep with thick white fleeces? How do you maintain enough tension to spin fibers together without breaking them? How do you prevent dyes from fading? How do you construct a loom that can weave complex patterns?

Some of the trickiest, however, are social: How do you finance a crop of silkworms or cotton, a new spinning mill, or a long-distance caravan? How do you record weaving patterns so someone else can duplicate them? How do you pay for textile shipments without physically sending currency? What do you do when the law forbids the cloth you want to make or use?

These questions arise from human universals. Human beings share the need for protection, the drive for status, and the pleasures of adornment. We are toolmaking, problem-solving animals and social, sensory creatures. Cloth embodies all these characteristics.

But universals are manifest in history only through particulars: the achievements of inventors, artists, and laborers, the longings of scientists and consumers, the initiative of explorers and entrepreneurs. The story of textiles encompasses beauty and genius, excess and cruelty, social hierarchies and subtle workarounds, peaceful trade and savage wars. Hidden in every piece of fabric are the actions of curious, clever, and desiring men and women, past and present, known and unknown, from every corner of the globe.

This heritage does not belong to a single nation, race, or culture, or to a single time or place. The story of textiles is not a male story or a female story, not a European, African, Asian, or American story. It is all of these, cumulative and shared—a *human* story, a tapestry woven from countless brilliant threads.

ACKNOWLEDGMENTS

My textile journey went from an idle concept to serious research in 2014, when Denita Sewell suggested I attend the Textile Society of America's biennial symposium being held at nearby UCLA. I owe her a big thank-you. I was fascinated by what I heard at the conference, particularly the papers from Marie-Louise Nosch on textile archaeology and Beverly Lemire on eighteenth-century trade. I also had stimulating conversations with them and other textile historians, including the great Betchen Barber.

From that beginning, I benefited from the enthusiasm and generosity of many textile scholars, businesspeople, and artisans. Some are featured in the book, and I thank them for sharing their work and their time. Others played an equally important role but do not necessarily appear by name. I want to thank them here.

Marie-Louise Nosch and Eva Andersson Strand hosted me on invaluable visits to the Centre for Textile Research at the University of Copenhagen, where Magdalena Öhrman, Jane Malcom-Davies, and Susanne Lervad furthered my textile education with conversations, references, and hands-on practice. Cherine Munkholt gave me early encouragement and introduced me to the work of Ellen Harlizius-Klück. Cécile Michel, whom I met at CTR, graciously shared her translations of Old Assyrian texts and answered endless questions about them.

John Styles, who was by happy coincidence visiting at the Huntington Library, gave me a veritable seminar on the literature about spinning and the Industrial Revolution. He, too, answered many email follow-ups. Claudio Zanier introduced me to Flavio Crippa, who not only arranged multiple tours but drove me back and forth from Milan. Helen Chang told me about textiles as money on the Silk Road. Deb McClintock helped me understand the Lao loom. While I was still crafting my proposal, Steve Gjerstad gave me a stack of useful economic history articles.

Diane Fagan Affleck and Karen Herbaugh shared their research on "neon colors" in nineteenth-century cotton prints and showed me sample books at the late lamented National Textile History Museum. Michelle McVicker selected garments from the archives of the Museum at FIT to demonstrate textiles before and after aniline dyes.

Meir Kohn shared his book manuscript and answered questions about economic institutions. Timur Kuran let me pick his brain about why textiles played such a central role in their evolution. Lining Yao shared her research on morphing materials. Tien Chiu responded to every email.

Gabriel Calzada invited me to the Universidad Francisco Marroquín in Guatemala City and arranged an extensive textile tour and writing retreat along with my visit. Pablo Velásquez, Isabel Moino, Lissa Hanckel, and Lisa Fitzpatrick were delightful hosts there.

My India trip would not have happened without the help of my friend Shikha Dalmia, my sister-in-law Jamie Inman, and their far-flung contacts. Shikha Banerjee took me on a whirlwind tour of Indian textiles at the Central Cottage Industries Emporium in New Delhi. Suresh Matur invited me to speak at Auto University in Surat, which set up a tour of the Laxmipati Saree plant and put me up at Suresh's beautiful hotel. Anju and Girish Sethi were wonderful local hosts, taking me shopping and opening doors to local factories.

Early on in my research, I realized that I would never understand looms unless I learned to use one. Trudy Sonia gave me introductory lessons, lent me a table loom, inspired me with her own beautiful work, and introduced me to the Southern California Handweavers' Guild. Along with cheering my book progress and early weaving efforts, the guild proved a valuable source of specialized reference materials. I particularly thank Chantal Hoareau and Amy Clark, who manage its three-thousand-item library, and Anna Zinsmeister, who lent me hard-to-find books on West African weaving.

When Postrel family library access failed, Brian Frye got me unpublished dissertations and other obscure publications. John Pearley Huffman made a special trip to the UCSB library to scan pages I needed in a hurry. Alex Knell served as my research elf, picking up and returning books from UCLA. Over the past few years, too many friends to name have sent me links to textile-related articles in the popular press. I particularly thank Cosmo Wenman, Dave Bernstein, Christine Whittington, and Richard Campbell.

Researching *The Fabric of Civilization* made me appreciate even more than usual the cornucopia of historic and scholarly materials available online, thanks to sites including Academia.edu, ResearchGate, and Google Books. The Internet Archive is an absolute treasure, containing copies of early editions found only in a handful of libraries. (I donate money to support the archive and encourage readers to do so as well.) The photos in this book reflect its contents, as well as the extraordinary number of public domain images now shared by many of the world's great museums.

Before there was a book, there was a 2015 *Aeon* article called "Losing the Thread," which found a home after Sonal Chokshi, to whom I owe great thanks, introduced me to Ross Andersen, who ably edited it. After reading that essay, Ben Platt invited me to submit a book proposal to Basic Books. I wasn't ready yet and by the time I was, Ben had left book publishing. But Leah Stecher brought the book to Basic after all. Things were going swimmingly when she, too, left the industry. What is normally an author's nightmare turned into an absolute delight, connecting me with Claire Potter, who proved an enthusiastic and insightful editor and enlisted Brandon Proia as a wise second pair of eyes. Throughout it all, my agents Sarah Chalfant, Jess Friedman, and Alex Christie have been unfailingly professional and supportive. Brynn Warriner shepherded the manuscript through production, Christina Palaia copyedited it, and Judy Kip did the index.

Amy Alkon, Joan Kron, Janet Levi, and Jonathan Rauch gave me feedback on early chapters. Leslie Watkins read and commented on each chapter as I wrote. Betchen Barber, Richard Campbell, Deirdre McCloskey, Grace Peng, and Leslie Rodier read the finished manuscript, giving me valuable feedback from their varied perspectives. Annabelle Gurwitch and Kathryn Bowers provided a quick turnaround read when I was struggling to rewrite the preface. Cameron Taylor-Brown, Deborah Graham, and Pat Sullivan brought fresh eyes and weavers' knowledge to the final proofreading.

I owe Lynn Scarlett a tremendous debt for lending me her Santa Barbara house as a writing hideout. Joan Kron put me up on many a New York trip, offering incomparable company as well as a beautiful place to stay.

When with great trepidation I asked David Shipley for a year off from writing columns for Bloomberg Opinion, he immediately said, "Of course." I thank him and my column editors, Jon Landsman, Katy Roberts, Toby Harshaw, James Gibney, Mike Nizza, Stacey Schick, and Brooke Sample. Thanks also to my friend and Bloomberg colleague, Adam Minter, whose book research on the global trade in secondhand clothes overlapped with my own; may we have many more conversations about textiles.

The research for this book was supported by a generous grant from the Alfred P. Sloan Foundation's program for the Public Understanding of Science, Technology, and Economics. I was greatly honored to receive this recognition and enormously grateful for the financial help. I thank Doron Weber and Ali Chunovic for their encouragement and assistance.

This book is dedicated to my parents, Sam and Sue Inman, not only because they are wonderful parents but also because *The Fabric of Civilization* reflects their intellectual influences: science and history from my father, art from my mother, writing and "making" from both. It is also dedicated to Steven Postrel, my best friend and true love, the indispensable man in my life and first reader of all my work—the steady warp to my weft.

GLOSSARY

abacist: arithmetic instructor in early modern Italy; also known as a *maestro d'abaco* or *abbachista*

alizarin: a red-orange dye compound

alum: a potassium or ammonium aluminum sulfate that was an important mordant

aniline: an alkaloid compound that became the major component of chemical dyes

Asantehene: Asante king

aulnager: government official in Britain who certified woolen cloth and collected taxes on it

bast fibers: stringy vascular tissue found inside the bark or stem of a plant and used to make yarn, string, or rope; bast fibers include flax, nettle, hemp, jute, lime, and willow.

bill of exchange: form letter telling an agent in another city to pay someone a certain amount

bistanclac: onomatopoeic Lyonnais word used to refer to looms using the Jacquard attachment

Bolinus brandaris: mollusk also known as the spiny dye murex that produces a red-violet dye

Bombyx mori: the mulberry silkworm responsible for cultivated silk

Bonwirehene: chief in charge of weaving in the Asante town of Bonwire, known for its kente cloth

bran water: an acid produced by soaking bran for several days, used in dyeing

brazilwood: dyestuff derived from the dense heartwood of certain tropical trees, sometimes called simply *brazil*; the country is named for it.

brocade: fabric, often using luxury threads, that incorporates supplemental weft to create a design

calcino: silkworm-killing disease investigated by Agostino Bassi in the early nineteenth century; also called *mal del segno*, *muscardine*, *calco*, or *calcinaccio*

calico: printed cotton fabric, originally from India, also known as *chintz* and *indienne*

charkha: Indian spindle wheel, particularly good for spinning cotton

chintz: printed cotton fabric, originally from India, also known as *calico* and *indienne*

chōnin: townspeople in Edo Japan, low-ranking commoners, including merchants

clothier: a cloth maker

coal tar: a sludge of assorted hydrocarbons left behind by the production of gas and coke from coal; it became a feedstuff for new chemical dyes.

cochineal: valuable red dyestuff made from the bodies of tiny parasitic insects that grow on the nopal, or prickly pear, cactus; this cultivated New World source replaced wild kermes.

corte: skirt worn as part of a traditional Maya ensemble, a long strip of cloth, usually woven on a floor loom, secured with a tightly cinched belt

distaff: pole for holding fibers prepared for spinning

double weave: cloth in which two layers are woven simultaneously

draft: diagram for weaving a particular pattern

drafting: first step of spinning thread, in which the spinner stretches out a bit of fiber still attached to a clean mass of wool, flax, or cotton

draper: fabric wholesaler

drawloom: large floor loom in which a weaver's assistant (sometimes called the *drawboy* or *drawgirl*) controls the raising of individual warp threads to create a brocade pattern

drop spindle: two-part mechanism for spinning thread, consisting of a stick with a spindle whorl at one end

dry exchange: payment of one bill of exchange with another one

factor: historically, as used here, an agent or middleman; in current usage, an entity that provides credit based on an apparel manufacturer's current invoices

faja: wide, tightly cinched belt worn as part of a traditional Maya ensemble

felt: cloth created by matting wet animal fibers together with friction

filament: a continuous extruded fiber, as in silk or synthetic fibers (opposed to *staple*)

framework knitting: an early form of mechanical knitting, invented in the sixteenth century

fustian: fabric using linen warp threads and cotton weft, produced in England as "cottons" before the Industrial Revolution

gabella: annual fee (officially a fine) that allowed early modern Florentines to wear otherwise forbidden luxury clothing

gauze: fabric structure in which pairs of warp threads are twisted together, with the weft inserted in the twist; also known as *leno* weave

Gossypium arboreum: the Old World cotton species native to the Indian subcontinent, often called tree cotton

***Gossypium barbadense*:** the long-fiber cotton species sometimes referred to as pima, Egyptian, or Sea Island cotton

***Gossypium herbaceum*:** one of two Old World cultivated cotton species, sometimes called Levant cotton, and the closest surviving descendant of the original African species from which all cotton fiber comes

***Gossypium hirsutum*:** the dominant species of cultivated cotton, originally from the Yucatan Peninsula

***grana*:** any of several prized red dyes derived from the bodies of tiny insects

hackling: running flax stems through combs to separate the long fibers from the short, fluffy tow

heddle: a loop of string or metal used to raise and lower warp threads

***Hexaplex trunculus*:** mollusk also known as the banded dye murex that produces several shades of purple dye

***huipil*:** Maya blouse made of pieces woven on a backstrap loom, usually with supplementary weft embellishment

ikat: cloth pattern created by tightly tying string around threads to block out areas before dyeing; ikat is distinguished by the slightly blurred appearance of the figures; if both warp and weft threads are tied, the cloth is called *double ikat*.

***iki*:** a style ideal developed in Edo Japan in which subtlety is highly valued

indican: indigo precursor found in plants

***indienne*:** printed cotton fabric, originally from India, also known as *calico* and *chintz*

***Indigofera tinctoria*:** South Asian legume known as "true indigo" in Europe

indigotin: insoluble blue pigment, otherwise known as *indigo*, that forms when indoxyl is exposed to oxygen

indoxyl: highly reactive, colorless compound produced when indigo leaves break down in water

Jacquard loom: loom with an attachment that enables it to automatically select individual warp threads to create a pattern; originally driven by mechanical punch cards, Jacquard looms now use computerized controllers.

jaspe: Guatemalan ikat

kente cloth: a West African strip cloth distinguished by its alternating blocks of weft-faced and warp-faced designs; the term is often used colloquially for cloth with patterns derived from kente designs.

kermes: valuable red dyestuff derived from the bodies of tiny insects that live on European oak trees, often called *grana*

knit stitch: the basic stitch in knitting. In hand knitting, a new loop is pulled down through the previous one.

lampas: a complex brocade structure that uses two warp systems and at least two wefts

leuco-indigo: soluble compound formed when indigotin breaks down in an alkaline environment. Sometimes called white indigo for its pale color.

madder: versatile red dye derived from the roots of *Rubia tinctorum*, a plant more commonly known as dyer's madder

magnetic core memory: early form of computer memory that used woven copper wires with a tiny ferrite bead at each cross, representing a single bit

mise-en-carte: large-scale graph-paper rendering of a brocade design

mordant: chemical, usually a metal salt, that causes dyes to securely attach to fibers

nålbinding: a way of creating cloth using a blunt needle to pass thread through loops wound around the thumb; unlike knitting, which uses only one end of the thread, *nålbinding* draws the entire length through each loop, using fairly short pieces that are fused together by felting when they run out. The need for felting limits *nålbinding* to animal fibers.

nasīj: silk brocade with gold threads used heavily by the Mongols, also known as *Tartar cloth* or *cloth of Tartary*

open tabby: a plain weave alternative to gauze

organzine: strong, multifilament silk threads used for warp

pébrine: a silkworm disease that devastated European sericulture in the nineteenth century, caused by a parasitic protozoan

pick: a weft row

plain weave: fabric structure created by alternating every other thread of the warp and weft; also known as *tabby*

polyploidy: the phenomenon in which an organism gets two copies of each parent's chromosomes rather than one of each; polyploidy is common in plants.

polyvoltine: reproducing multiple times a year (applies to insects, particularly silkworms)

purl stitch: a reversed knit stitch, which used in conjunction with the knit stitch creates ribs. In hand knitting, a new loop is pulled up through the previous one.

purpurin: a purple dye compound

quilling: winding spun thread or reeled silk onto bobbins

qilin: a dragon-like creature with cloven hoofs used as a pattern on high-status Ming dynasty robes

reed: comblike loom component that keeps warp threads separated and in order; may be combined with the beater

reeling: winding silk filaments off cocoons submerged in warm water

retting: soaking flax stems in water to break down the pectin that holds the bast fibers to the outer stem

rope memory: read-only computer memory used in the Apollo program

satin: smooth cloth structure woven with few intersections of warp and weft, arranged so they do not create twill-like diagonals

scarsella: regular messenger service established by Italian merchants beginning in the fourteenth century; literally, the messenger bag; plural *scarselle*

scutching: beating and scraping dried flax stems to separate the fiber from the straw

sericulture: the raising and harvesting of silkworms

shaft: bar that holds and lifts a row of heddles; often controlled by a lever or pedal

shed: space between raised and lowered warp threads allowing passage of shuttle containing weft

simple: collectively, the vertical cords that control the lifting of individual warp threads on a French drawloom; also known as the *semple*

spindle wheel: device that mechanized the first two stages of spinning, drawing out and twisting the fiber, using a belt drive

spindle whorl: small cone, disk, or sphere made of a hard material with a hole through the center; as part of a drop spindle, the whorl adds weight and increases the angular momentum.

staple: short fiber that must be spun to produce thread (opposed to *filament*)

stocking frame: an early mechanical device for knitting, developed in England in the sixteenth century and used primarily for stockings

Stramonita haemastoma: mollusk also known as the red-mouthed rock snail that produces a red-violet dye

strip cloth: fabric in which narrow strips are woven and sewn together into a larger cloth, particularly common in Africa; both the strips and final cloth are usually standard sizes.

sumptuary laws: laws restricting consumption, usually aimed at luxuries and often with regulations according to social class

supplementary weft: weft threads inserted over and under individually selected warp threads to add a design to cloth; they do not constitute structural components of the cloth.

tadeai: Japanese indigo, *Persicaria tinctoria*, also known as *Polygonum tinctorium* and dyer's knotweed

tapestry: weaving in which the structural weft threads create designs in discontinuous colors and completely obscure the warp

throwing: twisting silk filaments together

traje: traditional Maya ensemble worn in Guatemala

twill: fabric structure with diagonal ribs created by passing weft over or under multiple consecutive warp threads in a sequential, rather than an alternating, pattern; each new weft row shifts the pattern over one warp thread.

twisting: second step of spinning, in which individual fibers combine to form a continuous thread

usance: period after which a bill of exchange will be paid; originally developed to allow time to notify agents in distant cities that such a bill would be coming

vaðmál: standardized wool twill cloth used as currency in medieval Iceland

warp: strong threads held in tension and raised or lowered to create opening for weft to cross

weft: threads interlaced horizontally between raised and lowered warp to create fabric weave, often softer than warps; also called *woof*, especially in older writing

white tartar: a sediment produced in wine fermentation, used in dyeing

winding: final step of spinning, in which yarn is gathered into a skein that preserves its twist

woad: European indigo

NOTES

Preface: The Fabric of Civilization

1. Sylvia L. Horwitz, *The Find of a Lifetime: Sir Arthur Evans and the Discovery of Knossos* (New York: Viking, 1981); Arthur J. Evans, *Scripta Minoa: The Written Documents of Minoan Crete with Special Reference to the Archives of Knossos*, Vol. 1 (Oxford: Clarendon Press, 1909), 195–199; Marie-Louise Nosch, "What's in a Name? What's in a Sign? Writing Wool, Scripting Shirts, Lettering Linen, Wording Wool, Phrasing Pants, Typing Tunics," in *Verbal and Nonverbal Representation in Terminology Proceedings of the TOTh Workshop 2013, Copenhagen—8 November 2013*, ed. Peder Flemestad, Lotte Weilgaard Christensen, and Susanne Lervad (Copenhagen: SAXO, Københavns Universitet, 2016), 93–115; Marie-Louise Nosch, "From Texts to Textiles in the Aegean Bronze Age," in *Kosmos: Jewellery, Adornment and Textiles in the Aegean Bronze Age, Proceedings of the 13th International Aegean Conference/13e Rencontre égéenne internationale, University of Copenhagen, Danish National Research Foundation's Centre for Textile Research, 21–26 April 2010*, ed. Marie-Louise Nosch and Robert Laffineur (Liege: Petters Leuven, 2012), 46.

2. Clarke's "third law" states that any sufficiently advanced technology is indistinguishable from magic. See "Clarke's three laws," *Wikipedia*, last modified February 3, 2020, https://en.wikipedia.org/wiki/Clarke's_three_laws.

3. For a basic overview of the issues of defining civilization, see Cristian Violatti, "Civilization: Definition," *Ancient History Encyclopedia*, December 4, 2014, www.ancient.eu/civilization/. The definition cited here is from Mordecai M. Kaplan, *Judaism as a Civilization: Toward a Reconstruction of American-Jewish Life* (Philadelphia: Jewish Publication Society of America, 1981), 179.

4. Jerry Z. Muller, *Adam Smith in His Time and Ours: Designing the Decent Society* (New York: Free Press, 1993), 19.

5. Marie-Louise Nosch, "The Loom and the Ship in Ancient Greece: Shared Knowledge, Shared Terminology, Cross-Crafts, or Cognitive Maritime-Textile Archaeology," in *Weben und Gewebe in der Antike. Materialität—Repräsentation—Episteme—Metapoetik*, ed. Henriette Harich-Schwartzbauer (Oxford: Oxbow Books, 2015), 109–132. *Histology*, the study of tissues, comes from the same word, while *tissue* itself comes from *texere*.

6. *-teks*, www.etymonline.com/word/*teks-#etymonline_v_52573; Ellen Harlizius-Klück, "Arithmetics and Weaving from Penelope's Loom to Computing," Münchner Wissenschaftstage (poster), October 18–21, 2008; Patricia Marks Greenfield, *Weaving*

Generations Together: Evolving Creativity in the Maya of Chiapas (Santa Fe, NM: School of American Research Press, 2004), 151; *sutra*, www.etymonline.com/word/sutra; *tantra*, www.etymonline.com/word/tantra; Cheng Weiji, ed., *History of Textile Technology in Ancient China* (New York: Science Press, 1992), 2.

7. David Hume, "Of Refinement in the Arts," in *Essays, Moral, Political, and Literary*, ed. Eugene F. Miller (Indianapolis: Liberty Fund, 1987), 273, www.econlib.org/library /LFBooks/Hume/hmMPL25.html.

8. Italicized terms, such as *warp, weft, and strip cloths* here, may be found in the glossary.

Chapter One: Fiber

1. Elizabeth Wayland Barber, *Women's Work, the First 20,000 Years: Women, Cloth, and Society in Early Times* (New York: W. W. Norton, 1994), 45.

2. Karen Hardy, "Prehistoric String Theory: How Twisted Fibres Helped Shape the World," *Antiquity* 82, no. 316 (June 2008): 275. Nowadays Papua New Guineans tend to use commercially available yarn, which also offers a wider range of colors and textures, to make the versatile string bags called *bilums*. Barbara Andersen, "Style and Self-Making: String Bag Production in the Papua New Guinea Highlands," *Anthropology Today* 31, no. 5 (October 2015): 16–20.

3. M. L. Ryder, *Sheep & Man* (London: Gerald Duckworth & Co., 1983), 3–85; Melinda A. Zeder, "Domestication and Early Agriculture in the Mediterranean Basin: Origins, Diffusion, and Impact," *Proceedings of the National Academy of Sciences* 105, no. 33 (August 19, 2003): 11597–11604; Marie-Louise Nosch, "The Wool Age: Traditions and Innovations in Textile Production, Consumption and Administration in the Late Bronze Age Aegean" (paper presented at the Textile Society of America 2014 Biennial Symposium: New Directions: Examining the Past, Creating the Future, Los Angeles, CA, September 10–14, 2014).

4. In contemporary terminology, *linseed oil*, which is inedible because of the way it is processed, is sometimes distinguished from *flaxseed oil*, which is usually eaten as a nutritional supplement. In prehistoric times, there was no difference other than use, and even today *linseed oil* may refer to any oil made by pressing flaxseeds.

5. Ehud Weiss and Daniel Zohary, "The Neolithic Southwest Asian Founder Crops: Their Biology and Archaeobotany," Supplement, *Current Anthropology* 52, no. S4 (October 2011): S237–S254; Robin G. Allaby, Gregory W. Peterson, David Andrew Merriwether, and Yong-Bi Fu, "Evidence of the Domestication History of Flax (*Linum usitatissimum* L.) from Genetic Diversity of the *sad2* Locus," *Theoretical and Applied Genetics* 112, no. 1 (January 2006): 58–65. Whether plant alterations were conscious or unconscious is a matter of considerable scholarly debate because we can observe only the types of changes, not what the humans who made them were thinking. Although genetic analysis shows signs of selective breeding, dense planting also encourages flax to grow taller.

6. Samples of linen yarn were radiocarbon dated to 8,850 years old, plus or minus 90 years, and 9,210 years old, plus or minus 300 years. Twined and knotted fabric samples were dated to 8,500 years old, plus or minus 220 years, and 8,810 years old, plus or minus 120 years. Tamar Schick, "Cordage, Basketry, and Fabrics," in *Nahal Hemar Cave*, ed. Ofer Bar-Yosef and David Alon (Jerusalem: Israel Department of Antiquities and Museums, 1988), 31–38.

7. Jonathan Wendel, interviews with the author, September 21, 2017, and September 26, 2017, and email to the author, September 30, 2017; Susan V. Fisk, "Not Your Grandfather's Cotton," Crop Science Society of America, February 3, 2016, www.sciencedaily.com /releases/2016/02/160203150540.htm; Jonathan Wendel, "Phylogenetic History of *Gossypium*," video, www.eeob.iastate.edu/faculty/WendelJ/; J. F. Wendel, "New World Tetraploid Cottons Contain Old World Cytoplasm," *Proceedings of the National Academy of Science USA* 86, no. 11 (June 1989): 4132–4136; Jonathan F. Wendel and Corrinne E. Grover, "Taxonomy and Evolution of the Cotton Genus, Gossypium," in *Cotton*, ed. David D. Fang and Richard G. Percy (Madison, WI: American Society of Agronomy, 2015), 25–44, www .botanicaamazonica.wiki.br/labotam/lib/exe/fetch.php?media=bib:wendel2015.pdf; Jonathan F. Wendel, Paul D. Olson, and James McD. Stewart, "Genetic Diversity, Introgression, and Independent Domestication of Old World Cultivated Cotton," *American Journal of Botany* 76, no. 12 (December 1989): 1795–1806; C. L. Brubaker, F. M. Borland, and J. F. Wendel, "The Origin and Domestication of Cotton," in *Cotton: Origin, History, Technology, and Production*, ed. C. Wayne Smith and J. Tom Cothren (New York: John Wiley, 1999): 3–31.

8. Another possibility is that early-blooming cotton resisted pests, which was the case with the boll weevil in the southern United States.

9. Elizabeth Baker Brite and John M. Marston, "Environmental Change, Agricultural Innovation, and the Spread of Cotton Agriculture in the Old World," *Journal of Anthropological Archaeology* 32, no. 1 (March 2013): 39–53; Mac Marston, interview with the author, July 20, 2017; Liz Brite, interview with the author, June 30, 2017; Elizabeth Baker Brite, Gairatdin Khozhaniyazov, John M. Marston, Michelle Negus Cleary, and Fiona J. Kidd, "Kara-tepe, Karakalpakstan: Agropastoralism in a Central Eurasian Oasis in the 4th/5th Century A.D. Transition," *Journal of Field Archaeology* 42 (2017): 514–529, http://dx.doi .org/10.1080/00934690.2017.1365563.

10. Kim MacQuarrie, *The Last Days of the Incas* (New York: Simon & Schuster, 2007), 27–28, 58, 60; David Tollen, "Pre-Columbian Cotton Armor: Better than Steel," Pints of History, August 10, 2011, https://pintsofhistory.com/2011/08/10/mesoamerican-cotton -armor-better-than-steel/; Frances Berdan and Patricia Rieff Anawalt, *The Essential Codex Mendoza* (Berkeley: University of California Press, 1997), 186.

11. Sea Island cotton is a variety of *Gossypium barbadense*, the species originally cultivated in Peru; this species also includes long-staple pima cotton (and its trademarked variant Supima) and some so-called Egyptian cottons. More common "upland" cotton varieties are types of *Gossypium hirsutum*, the shorter-fiber species first cultivated on the Yucatan Peninsula. *G. hirsutum* currently accounts for about 90 percent of the world's commercial cotton, while *G. barbadense* makes up most of the rest. Whether generated randomly by nature or bred intentionally to enhance certain traits, a variety is a particular manifestation of the same species, as a poodle and a Great Dane are both dogs.

12. Jane Thompson-Stahr, *The Burling Books: Ancestors and Descendants of Edward and Grace Burling, Quakers (1600–2000)* (Baltimore: Gateway Press, 2001), 314–322; Robert Lowry and William H. McCardle, *The History of Mississippi for Use in Schools* (New York: University Publishing Company, 1900), 58–59.

13. John Hebron Moore, "Cotton Breeding in the Old South," *Agricultural History* 30, no. 3 (July 1956): 95–104; Alan L. Olmstead and Paul W. Rhode, *Creating Abundance: Biological Innovation and American Agricultural Development* (Cambridge: Cambridge

University Press, 2008), 98–133; O. L. May and K. E. Lege, "Development of the World Cotton Industry" in *Cotton: Origin, History, Technology, and Production*, ed. C. Wayne Smith and J. Tom Cothren (New York: John Wiley & Sons, 1999), 77–78.

14. Gavin Wright, *Slavery and American Economic Development* (Baton Rouge: Louisiana State University Press, 2006), 85; Dunbar Rowland, *The Official and Statistical Register of the State of Mississippi 1912* (Nashville, TN: Press of Brandon Printing, 1912), 135–136.

15. Edward E. Baptist, "'Stol' and Fetched Here': Enslaved Migration, Ex-slave Narratives, and Vernacular History," in *New Studies in the History of American Slavery*, ed. Edward E. Baptist and Stephanie M. H. Camp (Athens: University of Georgia Press, 2006), 243–274; Federal Writers' Project of the Works Progress Administration, *Slave Narratives: A Folk History of Slavery in the United States from Interviews with Former Slaves*, Vol. IX (Washington, DC: Library of Congress, 1941), 151–156, www.loc.gov/resource/mesn.090 /?sp=155.

16. In 1860, on the eve of the Civil War, the United States produced 4.56 million bales of cotton, a figure that dropped to 4.4 million in 1870 and jumped to 6.6 million in 1880. Between 1860 and 1870, the number of Southern cotton farms 40 hectares or smaller grew by 55 percent as former plantations were broken up and sold. Both black and white Southerners now worked as farm laborers, either on their own land, as sharecroppers, or as hired hands. The 1880s brought effective fertilization and new cotton breeds with larger bolls, making picking easier. May and Lege, "Development of the World Cotton Industry," 84–87; David J. Libby, *Slavery and Frontier Mississippi 1720–1835* (Jackson: University Press of Mississippi, 2004), 37–78. On the productivity effects and advantages to owners of property rights in slaves, see Wright, *Slavery and American Economic Development*, 83–122.

17. Cyrus McCormick, *The Century of the Reaper* (New York: Houghton Mifflin, 1931), 1–2, https://archive.org/details/centuryofthereap000250mbp/page/n23; Bonnie V. Winston, "Jo Anderson," *Richmond Times-Dispatch*, February 5, 2013, www.richmond.com /special-section/black-history/jo-anderson/article_277b0072-700a-11e2-bb3d-001a 4bcf6878.html.

18. Moore, "Cotton Breeding in the Old South," 99–101; M. W. Philips, "Cotton Seed," *Vicksburg (MS) Weekly Sentinel*, April 28, 1847, 1. For additional background on Philips, see Solon Robinson, *Solon Robinson, Pioneer and Agriculturalist: Selected Writings*, Vol. II, ed. Herbert Anthony Kellar (Indianapolis: Indianapolis Historical Bureau, 1936), 127–131.

19. Alan L. Olmstead and Paul W. Rhode, "Productivity Growth and the Regional Dynamics of Antebellum Southern Development" (NBER Working Paper No. 16494, Development of the American Economy, National Bureau of Economic Research, October 2010); Olmsted and Rhode, *Creating Abundance*, 98–133; Edward E. Baptist in *The Half Has Never Been Told: Slavery and the Making of American Capitalism* (New York: Basic Books, 2014), 111–144, argues that the productivity boost came from more effective methods of driving and torturing slaves, leading them to pick more efficiently. But the productivity increases are too large for that explanation, and the effects of new seeds are well documented. A better reading of the evidence is that plantation managers drove their slaves to pick as fast as seed technology allowed. John E. Murray, Alan L. Olmstead, Trevor D. Logan, Jonathan B. Pritchett, and Peter L. Rousseau, "Roundtable of Reviews for *The Half Has Never Been Told*," *Journal of Economic History*, September 2015, 919–931; "Baptism by Blood Cotton," Pseudoerasmus, September 12, 2014, https://pseudoerasmus.com/2014/09/12

/baptism-by-blood-cotton/, and "The Baptist Question Redux: Emancipation and Cotton Productivity," Pseudoerasmus, November 5, 2015, https://pseudoerasmus.com/2015/11/05 /bapredux/.

20. Yuxuan Gong, Li Li, Decai Gong, Hao Yin, and Juzhong Zhang, "Biomolecular Evidence of Silk from 8,500 Years Ago," *PLOS One* 11, no. 12 (December 12, 2016): e0168042, http://journals.plos.org/plosone/article?id=10.1371/journal.pone.0168042; "World's Oldest Silk Fabrics Discovered in Central China," Archaeology News Network, December 5, 2019, https://archaeologynewsnetwork.blogspot.com/2019/12/worlds-oldest-silk-fabrics -discovered.html; Dieter Kuhn, "Tracing a Chinese Legend: In Search of the Identity of the 'First Sericulturalist,'" *T'oung Pao*, nos. 4/5 (1984): 213–245.

21. Angela Yu-Yun Sheng, *Textile Use, Technology, and Change in Rural Textile Production in Song, China (960–1279)* (unpublished dissertation, University of Pennsylvania, 1990), 185–186.

22. Sheng, *Textile Use, Technology, and Change*, 23–40, 200–209.

23. J. R. Porter, "Agostino Bassi Bicentennial (1773–1973)," *Bacteriological Reviews* 37, no. 3 (September 1973): 284–288; Agostino Bassi, *Del Mal del Segno Calcinaccio o Moscardino* (Lodi: Dalla Tipografia Orcesi, 1835), 1–16, translations by the author; George H. Scherr, *Why Millions Died* (Lanham, MD: University Press of America, 2000), 78–98, 141–152; Seymore S. Block, "Historical Review," in *Disinfection, Sterilization, and Preservation*, 5th ed., ed. Seymour Stanton Block (Philadelphia: Lippincott Williams & Wilkins, 2001), 12.

24. Patrice Debré, *Louis Pasteur* (Baltimore: Johns Hopkins University Press, 2000), 177–218; Scherr, *Why Millions Died*, 110.

25. "The Cattle Disease in France," *Journal of the Society of the Arts*, March 30, 1866, 347; Omori Minoru, "Some Matters in the Study of von Siebold from the Past to the Present and New Materials Found in Relation to Siebold and His Works," *Historia scientiarum: International Journal of the History of Science Society of Japan*, no. 27 (September 1984): 96.

26. Tessa Morris-Suzuki, "Sericulture and the Origins of Japanese Industrialization," *Technology and Culture* 33, no. 1 (January 1992): 101–121.

27. Debin Ma, "The Modern Silk Road: The Global Raw-Silk Market, 1850–1930," *Journal of Economic History* 56, no. 2 (June 1996): 330–355, http://personal.lse.ac.uk/mad1 /ma_pdf_files/modern%20silk%20road.pdf; Debin Ma, "Why Japan, Not China, Was the First to Develop in East Asia: Lessons from Sericulture, 1850–1937," *Economic Development and Cultural Change* 52, no. 2 (January 2004): 369–394, http://personal.lse.ac.uk/mad1 /ma_pdf_files/edcc%20sericulture.pdf.

28. David Breslauer, Sue Levin, Dan Widmaier, and Ethan Mirsky, interviews with the author, February 19, 2016; Sue Levin, interview with the author, August 10, 2015; Jamie Bainbridge and Dan Widmaier, interviews with the author, February 8, 2017; Dan Widmaier, interviews with the author, March 21, 2018, and May 1, 2018.

29. Mary M. Brooks, "'Astonish the World with . . . Your New Fiber Mixture': Producing, Promoting, and Forgetting Man-Made Protein Fibers," in *The Age of Plastic: Ingenuity and Responsibility, Proceedings of the 2012 MCI Symposium*, ed. Odile Madden, A. Elena Charola, Kim Cullen, Cobb, Paula T. DePriest, and Robert J. Koestler (Washington, DC: Smithsonian Institution Scholarly Press, 2017), 36–50, https://smithsonian.figshare.com /articles/The_Age_of_Plastic_Ingenuity_and_Responsibility_Proceedings_of_the_2012

_MCI_Symposium_/9761735; National Dairy Products Corporation, "The Cow, the Milk-maid and the Chemist," www.jumpingfrog.com/images/epm10jun01/era8037b.jpg; British Pathé, "Making Wool from Milk (1937)," YouTube video, 1:24, April 13, 2014, www.youtube.com/watch?v=OyLnKz7uNMQ&feature=youtu.be; Michael Waters, "How Clothing Made from Milk Became the Height of Fashion in Mussolini's Italy," Atlas Obscura, July 28, 2017, www.atlasobscura.com/articles/lanital-milk-dress-qmilch; Maggie Koerth-Baker, "Aralac: The 'Wool' Made from Milk," Boing Boing, October 28, 2012, https://boing boing.net/2012/10/28/aralac-the-wool-made-from.html.

30. Dan Widmaier, interview with the author, December 16, 2019.

Chapter Two: Thread

1. *Yarn* and *thread* are synonyms and used interchangeably here. In the textile industry, *yarn* usually refers to all thread intended for weaving or knitting, whereas *thread* often refers specifically to sewing or embroidery yarn. *String* is usually reserved for cord used for tying or binding things, although all yarn and thread are also string.

2. Cordula Greve, "Shaping Reality through the Fictive: Images of Women Spinning in the Northern Renaissance," *RACAR: Revue d'art canadienne/Canadian Art Review* 19, nos. 1–2 (1992): 11–12.

3. Patricia Baines, *Spinning Wheels, Spinners and Spinning* (London: B. T. Batsford, 1977), 88–89.

4. Dominika Maja Kossowska-Janik, "Cotton and Wool: Textile Economy in the Serakhs Oasis during the Late Sasanian Period, the Case of Spindle Whorls from Gurukly Depe (Turkmenistan)," *Ethnobiology Letters* 7, no. 2 (2016): 107–116.

5. Elizabeth Barber, interview with the author, October 22, 2016; E. J. W. Barber, *Prehistoric Textiles: The Development of Cloth in the Neolithic and Bronze Ages with Special Reference to the Aegean* (Princeton, NJ: Princeton University Press, 1991), xxii.

6. Steven Vogel, *Why the Wheel Is Round: Muscles, Technology, and How We Make Things Move* (Chicago: University of Chicago Press, 2016), 205–208.

7. Sally Heaney, "From Spinning Wheels to Inner Peace," *Boston Globe*, May 23, 2004, http://archive.boston.com/news/local/articles/2004/05/23/from_spinning_wheels_to_inner_peace/.

8. Giovanni Fanelli, *Firenze: Architettura e città* (Florence: Vallecchi, 1973), 125–126; Celia Fiennes, *Through England on a Side Saddle in the Time of William and Mary* (London: Field & Tuer, 1888), 119; Yvonne Elet, "Seats of Power: The Outdoor Benches of Early Modern Florence," *Journal of the Society of Architectural Historians* 61, no. 4 (December 2002): 451, 466n; Sheilagh Ogilvie, *A Bitter Living: Women, Markets, and Social Capital in Early Modern Germany* (Oxford: Oxford University Press, 2003), 166; Hans Medick, "Village Spinning Bees: Sexual Culture and Free Time among Rural Youth in Early Modern Germany," in *Interest and Emotion: Essays on the Study of Family and Kinship*, ed. Hans Medick and David Warren Sabean (New York: Cambridge University Press, 1984), 317–339.

9. Tapan Raychaudhuri, Irfan Habib, and Dharma Kumar, eds., *The Cambridge Economic History of India: Volume 1, c. 1200–c. 1750* (Cambridge: Cambridge University Press, 1982), 78.

10. Rachel Rosenzweig, *Worshipping Aphrodite: Art and Cult in Classical Athens* (Ann Arbor: University of Michigan Press, 2004), 69; Marina Fischer, "Hetaira's Kalathos:

Prostitutes and the Textile Industry in Ancient Greece," *Ancient History Bulletin*, 2011, 9–28, www.academia.edu/12398486/Hetaira_s_Kalathos_Prostitutes_and_the_Textile _Industry_in_Ancient_Greece.

11. Linda A. Stone-Ferrier, *Images of Textiles: The Weave of Seventeenth-Century Dutch Art and Society* (Ann Arbor: UMI Research Press, 1985), 83–117; *Incogniti scriptoris nova Poemata, ante hac nunquam edita, Nieuwe Nederduytsche, Gedichten ende Raedtselen*, 1624, trans. Linda A. Stone-Ferrier, https://archive.org/details/ned-kbn-all-00000845-001.

12. Susan M. Spawn, "Hand Spinning and Cotton in the Aztec Empire, as Revealed by the *Codex Mendoza*," in *Silk Roads, Other Roads: Textile Society of America 8th Biennial Symposium*, September 26–28, 2002, Smith College, Northampton, MA, https:// digitalcommons.unl.edu/tsaconf/550/; Frances F. Berdan and Patricia Rieff Anawalt, *The Essential Codex Mendoza* (Berkeley: University of California Press, 1997), 158–164.

13. Constance Hoffman Berman, "Women's Work in Family, Village, and Town after 1000 CE: Contributions to Economic Growth?," *Journal of Women's History* 19, no. 3 (Fall 2007): 10–32.

14. This calculation assumes 1.75 yards of 60-inch-wide fabric, or a total of 3,780 square inches, with 62 warp threads and 40 weft threads per square inch.

15. Denim typically uses warp thread that runs 5,880 yards, or 3.34 miles, to the pound and weft thread that runs 5,040 yards, or 2.86 miles, to the pound. "Weaving with Denim Yarn," Textile Technology (blog), April 21, 2009, https://textiletechnology.word-press.com/2009/04/21/weaving-with-denim-yarn/; Cotton Incorporated, "An Iconic Staple," Lifestyle Monitor, August 10, 2016, http://lifestylemonitor.cottoninc.com/an-iconic-staple/; A. S. Bhalla, "Investment Allocation and Technological Choice—a Case of Cotton Spinning Techniques," *Economic Journal* 74, no. 295 (September 1964): 611–622, uses an estimate of 50 pounds of thread in 300 days, or 1 pound in 6 days.

16. A twin sheet is 72 inches by 102 inches, or 7,344 square inches. With 250 threads per square inch, that amounts to 1,836,000 inches, or 34.9 miles. A queen-size sheet is 92 inches by 102 inches, or 9,384 square inches. With 250 threads per square inch, that equals 2,346,200 inches, or 37 miles.

17. R. Patterson, "Wool Manufacture of Halifax," *Quarterly Journal of the Guild of Weavers, Spinners, and Dyers*, March 1958, 18–19. Patterson reports a spinning rate of 1 pound of wool per 12-hour day for medium-weight yarn. The calculation assumes 1,100 meters per pound. Merrick Posnansky, "Traditional Cloth from the Ewe Heartland," in *History, Design, and Craft in West African Strip-Woven Cloth: Papers Presented at a Symposium Organized by the National Museum of African Art, Smithsonian Institution, February 18–19, 1988* (Washington, DC: National Museum of African Art, 1992), 127–128. Posnansky records that it took a minimum of two days to spin a single skein of cotton and that a woman's cloth took a minimum of seventeen skeins. Dimensions vary, but a traditional Ewe woman's cloth is about 1 yard by 2 yards.

18. Ed Franquemont, "Andean Spinning . . . Slower by the Hour, Faster by the Week," in *Handspindle Treasury: Spinning Around the World* (Loveland, CO: Interweave Press, 2011), 13–14. Franquemont writes that it took "nearly 20 hours of work to spin a pound of yarn," which I have converted to 44 hours to spin a kilogram.

19. Eva Andersson, Linda Mårtensson, Marie-Louise B. Nosch, and Lorenz Rahmstorf, "New Research on Bronze Age Textile Production," *Bulletin of the Institute of Classical*

Studies 51 (2008): 171–174. The kilometer assumes 10 threads per square centimeter, or about 65 per square inch, which is significantly less than the 102 threads in a typical square inch of denim. That density is ignored in this calculation. The figure of 3,780 square inches used above is equivalent to 2.4 square meters.

20. Mary Harlow, "Textile Crafts and History," in *Traditional Textile Craft: An Intangible Heritage?*, 2nd ed., ed. Camilla Ebert, Sidsel Frisch, Mary Harlow, Eva Andersson Strand, and Lena Bjerregaard (Copenhagen: Centre for Textile Research, 2018), 133–139.

21. Eva Andersson Strand, "Segel och segelduksproduktion i arkeologisk kontext," in *Vikingetidens sejl: Festsrift tilegnet Erik Andersen*, ed. Morten Ravn, Lone Gebauer Thomsen, Eva Andersson Strand, and Henriette Lyngstrøm (Copenhagen: Saxo-Instituttet, 2016), 24; Eva Andersson Strand, "Tools and Textiles—Production and Organisation in Birka and Hedeby," in *Viking Settlements and Viking Society: Papers from the Proceedings of the Sixteenth Viking Congress*, ed. Svavar Sigmunddsson (Reykjavík: University of Iceland Press, 2011), 298–308; Lise Bender Jørgensen, "The Introduction of Sails to Scandinavia: Raw Materials, Labour and Land," *N-TAG TEN. Proceedings of the 10th Nordic TAG Conference at Stiklestad, Norway 2009* (Oxford: Archaeopress, 2012); Claire Eamer, "No Wool, No Vikings," *Hakai Magazine*, February 23, 2016, www.hakaimagazine.com/features /no-wool-no-vikings/.

22. Ragnheidur Bogadóttir, "Fleece: Imperial Metabolism in the Precolumbian Andes," in *Ecology and Power: Struggles over Land and Material Resources in the Past, Present and Future*, ed. Alf Hornborg, Brett Clark, and Kenneth Hermele (New York: Routledge, 2012), 87, 90.

23. Luca Mola, *The Silk Industry of Renaissance Venice* (Baltimore: Johns Hopkins University Press, 2003), 232–234.

24. Dieter Kuhn, "The Spindle-Wheel: A Chou Chinese Invention," *Early China* 5 (1979): 14–24, https://doi.org/10.1017/S0362502800006106.

25. Flavio Crippa, "Garlate e l'Industria Serica," Memorie e Tradizioni, Teleunica, January 28, 2015. Translation by the author based on transcript prepared by Dalila Cataldi, January 25, 2017. Flavio Crippa, interviews with the author, March 27 and 29, 2017; email to the author, May 14, 2018.

26. Carlo Poni, "The Circular Silk Mill: A Factory Before the Industrial Revolution in Early Modern Europe," in *History of Technology*, Vol. 21, ed. Graham Hollister-Short (London: Bloomsbury Academic, 1999), 65–85; Carlo Poni, "Standards, Trust and Civil Discourse: Measuring the Thickness and Quality of Silk Thread," in *History of Technology*, Vol. 23, ed. Ian Inkster (London, Bloomsbury Academic, 2001), 1–16; Giuseppe Chicco, "L'innovazione Tecnologica nella Lavorazione della Seta in Piedmonte a Metà Seicento," *Studi Storici*, January–March 1992, 195–215.

27. Roberto Davini, "A Global Supremacy: The Worldwide Hegemony of the Piedmontese Reeling Technologies, 1720s–1830s," in *History of Technology*, Vol. 32, ed. Ian Inkster (London, Bloomsbury Academic, 2014), 87–103; Claudio Zanier, "Le Donne e il Ciclo della Seta," in *Percorsi di Lavoro e Progetti di Vita Femminili*, ed. Laura Savelli and Alessandra Martinelli (Pisa: Felici Editore), 25–46; Claudio Zanier, emails to the author, November 17 and 29, 2016.

28. John Styles, interview with the author, May 16, 2018.

29. Arthur Young, *A Six Months Tour through the North of England*, 2nd ed. (London: W. Strahan, 1771), 3:163–164, 3:187–202; Arthur Young, *A Six Months Tour through the*

North of England (London: W. Strahan, 1770), 4:582. Spinners were paid on a piecework basis and didn't necessarily spin all day, but Young consistently asked about weekly earnings for full-time work. Craig Muldrew, "'Th'ancient Distaff' and 'Whirling Spindle': Measuring the Contribution of Spinning to Household Earning and the National Economy in England, 1550–1770," *Economic History Review* 65, no. 2 (2012): 498–526.

30. Deborah Valenze, *The First Industrial Woman* (New York: Oxford University Press, 1995), 72–73.

31. John James, *History of the Worsted Manufacture in England, from the Earliest Times* (London: Longman, Brown, Green, Longmans & Roberts, 1857), 280–281; James Bischoff, *Woollen and Worsted Manufacturers and the Natural and Commercial History of Sheep, from the Earliest Records to the Present Period* (London: Smith, Elder & Co., 1862), 185.

32. Beverly Lemire, *Cotton* (London: Bloomsbury, 2011), 78–79.

33. John Styles, "Fashion, Textiles and the Origins of the Industrial Revolution," *East Asian Journal of British History*, no. 5 (March 2016): 161–189; Jeremy Swan, "Derby Silk Mill," *University of Derby Magazine*, November 27, 2016, 32–34, https://issuu .com/university_of_derby/docs/university_of_derby_magazine_-_nove and https://blog .derby.ac.uk/2016/11/derby-silk-mill/; "John Lombe: Silk Weaver," Derby Blue Plaques, http://derbyblueplaques.co.uk/john-lombe/. Financial information from Clive Emsley, Tim Hitchcock, and Robert Shoemaker, "London History—Currency, Coinage and the Cost of Living," Old Bailey Proceedings Online, www.oldbaileyonline.org/static/Coinage .jsp.

34. Styles, "Fashion, Textiles and the Origins of the Industrial Revolution," and interview with the author, May 16, 2018; R. S. Fitton, *The Arkwrights: Spinners of Fortune* (Manchester, UK: Manchester University Press, 1989), 8–17.

35. Lemire, *Cotton*, 80–83.

36. Deirdre Nansen McCloskey, *Bourgeois Equality: How Ideas, Not Capital, Transformed the World* (Chicago: University of Chicago Press, 2016), 8.

37. David Sasso, interviews with the author, May 22–23, 2018. The calculation is based on spinning four pounds a week, which is taken from Jane Humphries and Benjamin Schneider, "Spinning the Industrial Revolution," *Economic History Review* 72, no. 1 (May 23, 2018), https://doi.org/10.1111/ehr.12693.

Chapter Three: Cloth

1. Gillian Vogelsang-Eastwood, Intensive Textile Course, Textile Research Centre, September 15, 2015.

2. Kalliope Sarri, "Neolithic Textiles in the Aegean" (presentation at Centre for Textile Research, Copenhagen, September 22, 2015); Kalliope Sarri, "In the Mind of Early Weavers: Perceptions of Geometry, Metrology and Value in the Neolithic Aegean" (workshop abstract, "Textile Workers: Skills, Labour and Status of Textile Craftspeople between Prehistoric Aegean and Ancient Near East," Tenth International Congress on the Archaeology of the Ancient Near East, Vienna, April 25, 2016), https://ku-dk.academia.edu /KalliopeSarri.

3. sarah-marie belcastro, "Every Topological Surface Can Be Knit: A Proof," *Journal of Mathematics and the Arts* 3 (June 2009): 67–83; sarah-marie belcastro and Carolyn Yackel, "About Knitting . . . ," *Math Horizons* 14 (November 2006): 24–27, 39.

4. Carrie Brezine, "Algorithms and Automation: The Production of Mathematics and Textiles," in *The Oxford Handbook of the History of Mathematics*, ed. Eleanor Robson and Jacqueline Stedall (Oxford: Oxford University Press, 2009), 490.

5. Victor H. Mair, "Ancient Mummies of the Tarim Basin," *Expedition*, Fall 2016, 25–29, www.penn.museum/documents/publications/expedition/PDFs/58-2/tarim_basin.pdf.

6. O. Soffer, J. M. Adovasio, and D. C. Hyland, "The 'Venus' figurines: Textiles, Basketry, Gender, and Status in the Upper Paleolithic," *Current Anthropology* 41, no. 4 (August–October 2000): 511–537.

7. Jennifer Moore, "Doubleweaving with Jennifer Moore," *Weave* podcast, May 24, 2019, Episode 65, 30:30, www.gistyarn.com/blogs/podcast/episode-65-doubleweaving-with-jennifer-moore.

8. Technically *satin* is warp-faced and *sateen* is weft-faced, but the term *satin* is usually applied to the basic structure, for which the principle is the same.

9. Tien Chiu, interview with the author, July 11, 2018.

10. Ada Augusta, Countess of Lovelace, "Notes upon the Memoir by the Translator," in L. F. Menabrea, "Sketch of the Analytical Engine Invented by Charles Babbage," *Bibliothèque Universelle de Genève*, no. 82 (October 1842), www.fourmilab.ch/babbage/sketch.html.

11. E. M. Franquemont and C. R. Franquemont, "Tanka, Chongo, Kutij: Structure of the World through Cloth," in *Symmetry Comes of Age: The Role of Pattern in Culture*, ed. Dorothy K. Washburn and Donald W. Crowe (Seattle: University of Washington Press, 2004), 177–214; Edward Franquemont and Christine Franquemont, "Learning to Weave in Chinchero," *Textile Museum Journal* 26 (1987): 55–78; Ann Peters, "Ed Franquemont (February 17, 1945–March 11, 2003)," *Andean Past* 8 (2007): art. 10, http://digitalcommons.library.umaine.edu/andean_past/vol8/iss1/10.

12. Lynn Arthur Steen, "The Science of Patterns," *Science* 240, no. 4852 (April 29, 1988): 611–616.

13. Euclid's *Elements*, https://mathcs.clarku.edu/~djoyce/java/elements/elements.html.

14. Ellen Harlizius-Klück, interview with the author, August 7, 2018, and emails to the author, August 28, August 29, September 13, 2018; Ellen Harlizius-Klück, "Arithmetics and Weaving: From Penelope's Loom to Computing," Münchner Wissenschaftstage, October 18–21, 2008, www.academia.edu/8483352/Arithmetic_and_Weaving._From_Penelopes_Loom_to_Computing; Ellen Harlizius-Klück and Giovanni Fanfani, "(B)orders in Ancient Weaving and Archaic Greek Poetry," in *Spinning Fates and the Song of the Loom: The Use of Textiles, Clothing and Cloth Production as Metaphor, Symbol and Narrative Device in Greek and Latin Literature*, ed. Giovanni Fanfani, Mary Harlow, and Marie-Louise Nosch (Oxford: Oxbow Books, 2016), 61–99.

15. Rather than a loom, per se, the border was probably created using tablet weaving, in which warp threads run through holes at the corners of a square card—back then of wood or clay, today of cardboard or plastic. The weaver stretches the threads tight by tying them to posts, and the top and bottom of the cards create the shed. By turning the cards either all at once or selectively, the weaver locks in the weft threads and can create patterns using different colors. The more cards, the more complex the pattern can be.

16. Jane McIntosh Snyder, "The Web of Song: Weaving Imagery in Homer and the Lyric Poets," *Classical Journal* 76, no. 3 (February/March 1981): 193–196; Plato, *The Being*

of the Beautiful: Plato's Thaetetus, Sophist, and Statesman, trans. with commentary by Seth Bernadete (Chicago: University of Chicago Press, 1984), III.31–III.33, III.66–III.67, III.107–III.113.

17. Sheramy D. Bundrick, "The Fabric of the City: Imaging Textile Production in Classical Athens," *Hesperia: The Journal of the American School of Classical Studies at Athens* 77, no. 2 (April–June 2008): 283–334; Monica Bowen, "Two Panathenaic Peploi: A Robe and a Tapestry," Alberti's Window (blog), June 28, 2017, http://albertis-window.com/2017/06 /two-panathenaic-peploi/; Evy Johanne Håland, "Athena's Peplos: Weaving as a Core Female Activity in Ancient and Modern Greece," *Cosmos* 20 (2004): 155–182, www.academia .edu/2167145/Athena_s_Peplos_Weaving_as_a_Core_Female_Activity_in_Ancient_and _Modern_Greece; E. J. W. Barber, "The Peplos of Athena," in *Goddess and Polis: The Panathenaic Festival in Ancient Athens*, ed. Jenifer Neils (Princeton, NJ: Princeton University Press, 1992), 103–117.

18. Donald E. Knuth, *Art of Computer Programming, Volume 2: Seminumerical Algorithms* (Boston: Addison-Wesley Professional, 2014), 294.

19. Anthony Tuck, "Singing the Rug: Patterned Textiles and the Origins of Indo-European Metrical Poetry," *American Journal of Archaeology* 110, no. 4 (October 2006): 539–550; John Kimberly Mumford, *Oriental Rugs* (New York: Scribner, 1921), 25. For examples of war rugs, which originated during the Soviet occupation of Afghanistan, see warrug.com. Mimi Kirk, "Rug-of-War," *Smithsonian*, February 4, 2008, www.smithsonianmag.com /arts-culture/rug-of-war-19377583/. For an example of rug weavers singing their patterns, see Roots Revival, "Pattern Singing in Iran—'The Woven Sounds'—Demo Documentary by Mehdi Aminian," YouTube video, 10:00, March 15, 2019, www.youtube.com /watch?v=vhgHJ6xiau8&feature=youtu.be.

20. Eric Boudot and Chris Buckley, *The Roots of Asian Weaving: The He Haiyan Collection of Textiles and Looms from Southwest China* (Oxford: Oxbow Books, 2015), 165–169.

21. Malika Kraamer, "Ghanaian Interweaving in the Nineteenth Century: A New Perspective on Ewe and Asante Textile History," *African Arts*, Winter 2006, 44. For more on this topic, see Chapter 6.

22. "Ancestral Textile Replicas: Recreating the Past, Weaving the Present, Inspiring the Future" (exhibition, Museum and Catacombs of San Francisco de Asís of the City of Cusco, November 2017).

23. Nancy Arthur Hoskins, "Woven Patterns on Tutankhamun Textiles," *Journal of the American Research Center in Egypt* 47 (2011): 199–215, www.jstor.org/stable/24555392.

24. Richard Rutt, *A History of Hand Knitting* (London: B. T. Batsford, 1987), 4–5, 8–9, 23, 32–39. Native peoples in a region encompassing parts of today's Venezuela, Guyana, and Brazil separately developed their own form of knitting. Rutt notes that the words used to describe knitting appear no earlier than the early modern period and in many places were borrowed either from other countries—Russian adapted the French term *tricot*, for instance—or other textile crafts. "The contrast with words of 'weaving' is striking," he writes. "In most languages there is a precise, ancient and well developed vocabulary for weaving. Weaving is older than history. The apparently simple process of knitting turns out to be much less ancient."

25. Anne DesMoines, interview with the author, December 8, 2019; Anne DesMoines, "Eleanora of Toledo Stockings," www.ravelry.com/patterns/library/eleonora-di

-toledo-stockings. DesMoines says her published pattern is somewhat simplified compared to her exact reproduction, which includes more complicated shaping.

26. Although the examples of the cloth survived, sometime after the Spanish conquest, Andean weavers forgot an image-making technique called double weave pickup, which they'd used for thousands of years. In 2012, the Center for Traditional Textiles of Cusco recruited Jennifer Moore, an American double weave artist and teacher, to reintroduce the technique to master weavers, who could pass it on to others. An English speaker accustomed to floor looms, she spent a year preparing. Jennifer Moore, "Teaching in Peru," www .doubleweaver.com/peru.html.

27. Patricia Hilts, *The Weavers Art Revealed: Facsimile, Translation, and Study of the First Two Published Books on Weaving: Marx Ziegler's Weber Kunst und Bild Buch (1677) and Nathaniel Lumscher's Neu eingerichtetes Weber Kunst und Bild Buch (1708)*, Vol. I (Winnipeg, Canada: Charles Babbage Research Centre, 1990), 9–56, 97–109.

28. Joel Mokyr, *The Gifts of Athena: Historical Origins of the Knowledge Economy* (Princeton, NJ: Princeton University Press, 2002), 28–77.

29. Ellen Harlizius-Klück, "Weaving as Binary Art and the Algebra of Patterns," *Textile* 1, no. 2 (April 2017): 176–197.

30. If the ground and supplementary weft were the same color, the resulting fabric would be a *damask*.

31. Demonstration at "A World of Looms," China National Silk Museum, Hangzhou, June 1–4, 2018. Before inexpensive nylon string, thin bamboo rods were used, as they still are for simple patterns. Deb McClintock, "The Lao Khao Tam Huuk, One of the Foundations of Lao Pattern Weaving," Looms of Southeast Asia, January 31, 2017, https://simplelooms.com/2017/01/31/the-lao-khao-tam-huuk-one-of-the-foundations-of-lao -pattern-weaving/; Deb McClintock, interview with the author, October 18, 2018; Wendy Garrity, "Laos: Making a New Pattern Heddle," Textile Trails, https://textiletrails.com .au/2015/05/22/laos-making-a-new-pattern-heddle/.

32. E. J. W. Barber, *Prehistoric Textiles: The Development of Cloth in the Neolithic and Bronze Ages with Special Reference to the Aegean* (Princeton, NJ: Princeton University Press, 1991), 137–140.

33. Boudot and Buckley, *The Roots of Asian Weaving*, 180–185, 292–307, 314–327; Chris Buckley, email to the author, October 21, 2018.

34. Boudot and Buckley, *The Roots of Asian Weaving*, 422–426.

35. Boudot and Buckley, *The Roots of Asian Weaving*, 40–44.

36. Claire Berthommier, "The History of Silk Industry in Lyon" (presentation at the Dialogue with Silk between Europe and Asia: History, Technology and Art Conference, Lyon, November 30, 2017).

37. Daryl M. Hafter, "Philippe de Lasalle: From *Mise-en-carte* to Industrial Design," *Winterthur Portfolio*, 1977, 139–164; Lesley Ellis Miller, "The Marriage of Art and Commerce: Philippe de Lasalle's Success in Silk," *Art History* 28, no. 2 (April 2005): 200–222; Berthommier, "The History of Silk Industry in Lyon"; Rémi Labrusse, "Interview with Jean-Paul Leclercq," trans. Trista Selous, *Perspective*, 2016, https://journals.openedition .org/perspective/6674; Guy Scherrer, "Weaving Figured Textiles: Before the Jacquard Loom and After" (presentation at Conference on World Looms, China National Silk Museum, Hangzhou, May 31, 2018), YouTube video, 18:27, June 29, 2018, www.youtube.com

/watch?v=DLAzP53l-D4; Alfred Barlow, *The History and Principles of Weaving by Hand and by Power* (London: Sampson Low, Marston, Searle, & Rivington, 1878), 128–139.

38. Metropolitan Museum of Art, "Joseph Marie Jacquard, 1839," www.metmuseum .org/art/collection/search/222531; Charles Babbage, *Passages in the Life of a Philosopher* (London: Longman, Green, Longman, Roberts & Green, 1864), 169–170.

39. Rev. R. Willis, "On Machinery and Woven Fabrics," in *Report on the Paris Exhibition of 1855*, Part II, 150, quoted in Barlow, *The History and Principles of Weaving by Hand and by Power*, 140–141.

40. James Payton, "Weaving," in *Encyclopaedia Britannica*, 9th ed., Vol. 24, ed. Spencer Baynes and W. Robertson Smith (Akron: Werner Co., 1905), 491–492, http://bit.ly /2AB1JVU; Victoria and Albert Museum, "How Was It Made? Jacquard Weaving," YouTube video, 3:34, October 8, 2015, www.youtube.com/watch?v=K6NgMNvK52A; T. F. Bell, *Jacquard Looms: Harness Weaving* (Read Books, 2010), Kindle edition reprint of T. F. Bell, *Jacquard Weaving and Designing* (London: Longmans, Green, & Co., 1895).

41. James Essinger, *Jacquard's Web: How a Hand-Loom Led to the Birth of the Information Age* (Oxford: Oxford University Press, 2007), 35–38; Jeremy Norman, "The Most Famous Image in the Early History of Computing," HistoryofInformation.com, www .historyofinformation.com/expanded.php?id=2245; Yiva Fernaeus, Martin Jonsson, and Jakob Tholander, "Revisiting the Jacquard Loom: Threads of History and Current Patterns in HCI," *CHI '12: Proceedings of the SIGCHI Conference on Human Factors in Computing Systems*, May 5–10, 2012, 1593–1602, https://dl.acm.org/citation.cfm?doid=2207676.220 8280.

42. Gadagne Musées, "The Jacquard Loom," inv.50.144, Room 21: Social Laboratory—19th C., www.gadagne.musees.lyon.fr/index.php/history_en/content/download /2939/27413/file/zoom_jacquard_eng.pdf; Barlow, *The History and Principles of Weaving by Hand and by Power*, 144–147; Charles Sabel and Jonathan Zeitlin, "Historical Alternatives to Mass Production: Politics, Markets and Technology in Nineteenth-Century Industrialization," *Past and Present*, no. 108 (August 1985): 133–176; Anna Bezanson, "The Early Use of the Term Industrial Revolution," *Quarterly Journal of Economics* 36, no. 2 (February 1922): 343–349; Ronald Aminzade, "Reinterpreting Capitalist Industrialization: A Study of Nineteenth-Century France," *Social History* 9, no. 3 (October 1984): 329–350. Although they eventually accepted the new technology, Lyonnais workers didn't remain quiescent. The uprisings of the *canuts*, or silk workers, in 1831 and 1834 are milestones in French labor and political history.

43. James Burke, "Connections Episode 4: Faith in Numbers," https://archive.org /details/james-burke-connections_s01e04; F. G. Heath, "The Origins of the Binary Code," *Scientific American*, August 1972, 76–83.

44. Robin Kang, interview with the author, January 9, 2018; Rolfe Bozier, "How Magnetic Core Memory Works," Rolfe Bozier (blog), August 10, 2015, https://rolfebozier.com /archives/113; Stephen H. Kaisler, *Birthing the Computer: From Drums to Cores* (Newcastle upon Tyne, UK: Cambridge Scholars Publishing, 2017), 73–75; Daniela K. Rosner, Samantha Shorey, Brock R. Craft, and Helen Remick, "Making Core Memory: Design Inquiry into Gendered Legacies of Engineering and Craftwork," *Proceedings of the 2018 CHI Conference on Human Factors in Computing Systems (CHI '18)*, paper 531, https://faculty.washington .edu/dkrosner/files/CHI-2018-Core-Memory.pdf.

45. Core memory was RAM (random-access memory), whereas rope memory was ROM (read-only memory).

46. David A. Mindell, *Digital Apollo: Human and Machine in Spaceflight* (Cambridge, MA: MIT Press, 2008), 154–157; David Mindell interview in *Moon Machines: The Navigation Computer*, YouTube video, Nick Davidson and Christopher Riley (directors), 2008, 44:21, www.youtube.com/watch?v=9YA7X5we8ng; Robert McMillan, "Her Code Got Humans on the Moon—and Invented Software Itself," *Wired*, October 13, 2015, www.wired.com/2015/10/margaret-hamilton-nasa-apollo/.

47. Frederick Dill, quoted in Rosner et al., "Making Core Memory."

48. Fiber Year Consulting, *The Fiber Year 2017* (Fiber Year, 2017), www.groz-beckert.com/mm/media/web/9_messen/bilder/veranstaltungen_1/2017_6/the_fabric_year/Fabric_Year_2017_Handout_EN.pdf. In 2016, knitting accounted for 57 percent of worldwide fabric sales by weight, compared to 32 percent for weaving, with knitting sales growing at 5 percent a year, compared to 2 percent for weaving.

49. Stanley Chapman, *Hosiery and Knitwear: Four Centuries of Small-Scale Industry in Britain c. 1589–2000* (Oxford: Oxford University Press, 2002), xx–27, 66–67. Chapman argues convincingly that in the Midlands, where framework knitting flourished, ordinary blacksmiths—not, say, silversmiths or clockmakers—developed the skills to make the required parts. Local smiths were known for their fine work, and there are no records of these other crafts. Pseudoerasmus, "The Calico Acts: Was British Cotton Made Possible by Infant Industry Protection from Indian Competition?" Pseudoerasmus (blog), January 5, 2017, https://pseudoerasmus.com/2017/01/05/ca/. For a video explaining how the stocking frame worked, see https://youtu.be/WdVDoLqg2_c.

50. Vidya Narayanan and Jim McCann, interviews with the author, August 6, 2019; Vidya Narayanan, interview with the author, December 11, 2019, and email to the author, December 11, 2019; Michael Seiz, interviews with the author, December 10, 2019, and December 11, 2019; Randall Harward, interview with the author, November 12, 2019; Vidya Narayanan, Kui Wu, Cem Yuksel, and James McCann, "Visual Knitting Machine Programming," *ACM Transactions on Graphics* 38, no. 4 (July 2019), https://textiles-lab.github.io/publications/2019-visualknit/.

Chapter Four: Dye

1. Tom D. Dillehay, "Relevance," in *Where the Land Meets the Sea: Fourteen Millennia of Human History at Huaca Prieta, Peru*, ed. Tom D. Dillehay (Austin: University of Texas Press, 2017), 3–28; Jeffrey Splitstoser, "Twined and Woven Artifacts: Part 1: Textiles," in *Where the Land Meets the Sea*, 458–524; Jeffrey C. Splitstoser, Tom D. Dillehay, Jan Wouters, and Ana Claro, "Early Pre-Hispanic Use of Indigo Blue in Peru," *Science Advances* 2, no. 9 (September 14, 2016), http://advances.sciencemag.org/content/2/9/e1501623.full. In addition to blue, the fragments also have stripes made by plying cotton with the bright-white fibers of a milkweed-like shrub.

2. Dominique Cardon, *Natural Dyes: Sources, Tradition Technology and Science*, trans. Caroline Higgett (London: Archetype, 2007), 1, 51, 167–176, 242–250, 360, 409–411.

3. Zvi C. Koren, "Modern Chemistry of the Ancient Chemical Processing of Organic Dyes and Pigments," in *Chemical Technology in Antiquity*, ed. Seth C. Rasmussen, ACS Symposium Series (Washington, DC: American Chemical Society, 2015), 197; Cardon, *Natural Dyes*, 51.

4. John Marshall, *Singing the Blues: Soulful Dyeing for All Eternity* (Covelo, CA: Saint Titus Press, 2018), 11–12. Some indigo plants, including woad, contain other indoxyl precursors as well.

5. Plant-derived dyes can look richer than synthetic colors because they include more than one color compound.

6. Deborah Netburn, "6,000-Year-Old Fabric Reveals Peruvians Were Dyeing Textiles with Indigo Long Before Egyptians," *Los Angeles Times*, September 16, 2016, www.latimes .com/science/sciencenow/la-sci-sn-oldest-indigo-dye-20160915-snap-story.html.

7. A highly acidic solution would also work, but historically indigo dyers have used alkaline additives. Cardon, *Natural Dyes*, 336–353.

8. Jenny Balfour-Paul, *Indigo: Egyptian Mummies to Blue Jeans* (Buffalo, NY: Firefly Books, 2011), 121–122.

9. Balfour-Paul, *Indigo*, 41–42.

10. Alyssa Harad, "Blue Monday: Adventures in Indigo," Alyssa Harad, November 12, 2012, https://alyssaharad.com/2012/11/blue-monday-adventures-in-indigo/; Cardon, *Natural Dyes*, 369; Graham Keegan workshop, December 14, 2018.

11. Balfour-Paul, *Indigo*, 9, 13.

12. Cardon, *Natural Dyes*, 51, 336–353.

13. Graham Keegan, interview with the author, December 14, 2018.

14. Cardon, *Natural Dyes*, 571; Mark Cartwright, "Tyrian Purple," *Ancient History Encyclopedia*, July 21, 2016, www.ancient.eu/Tyrian_Purple; Mark Cartwright, "Melqart," *Ancient History Encyclopedia*, May 6, 2016, www.ancient.eu/Melqart/.

15. Cardon, *Natural Dyes*, 551–586; Zvi C. Koren, "New Chemical Insights into the Ancient Molluskan Purple Dyeing Process," in *Archaeological Chemistry VIII*, ed. R. Armitage et al. (Washington, DC: American Chemical Society, 2013), chap. 3, 43–67.

16. Inge Boesken Kanold, "Dyeing Wool and Sea Silk with Purple Pigment from *Hexaplex trunculus*," in *Treasures from the Sea: Purple Dye and Sea Silk*, ed. Enegren Hedvig Landenius and Meo Francesco (Oxford: Oxbow Books, 2017), 67–72; Cardon, *Natural Dyes*, 559–562; Koren, "New Chemical Insights."

17. Brendan Burke, *From Minos to Midas: Ancient Cloth Production in the Aegean and in Anatolia* (Oxford: Oxbow Books, 2010), Kindle locations 863–867. In a December 2, 2019, email to the author, Burke elaborates: "The idea of the cannibalism comes up because, IF they are kept in a fish tank, but temporarily denied access to food sources, they might start to eat each other. (I've always thought anyone who was tending to them would likely know that they need to feed these things—but perhaps not.) This has been the explanation for why some excavated deposits of snail shells connected to purple dyeing show the bored holes—but only some among large deposits of snails. So yes, the bore holes are a problem and I suspect the larger scale/professionalized production centers would learn that and they would not show up archaeologically quite as frequently as in the smaller scale workshops. The bored holes further suggest that whoever was the snail-keeper was not doing a good job feeding them."

18. Cardon, *Natural Dyes*, 559–562; Koren, "New Chemical Insights"; Zvi C. Koren, "Chromatographic Investigations of Purple Archaeological Bio-Material Pigments Used as Biblical Dyes," *MRS Proceedings* 1374 (January 2012): 29–47, https://doi.org/10.1557 /opl.2012.1376.

19. I am using the term Technicolor colloquially. The movie was actually filmed using a different color technology.

20. Meyer Reinhold, *History of Purple as a Status Symbol in Antiquity* (Brussels: Revue d'Études Latines, 1970), 17; Pliny, *Natural History*, Vol. III, Book IX, sec. 50, trans. Harris Rackham, Loeb Classical Library (Cambridge, MA: Harvard University Press, 1947), 247–259, https://archive.org/stream/naturalhistory03plinuoft#page/n7/mode/2up; Cassiodorus, "King Theodoric to Theon, Vir Sublimis," *The Letters of Cassiodorus*, Book I, trans. Thomas Hodgkin (London: Henry Frowde, 1886), 143–144, www.gutenberg.org/files/18590/18590-h/18590-h.htm; Martial, "On the Stolen Cloak of Crispinus," in *Epigrams*, Book 8, Bohn's Classical Library, 1897, adapted by Roger Pearse, 2008, www.tertullian.org/fathers/martial_epigrams_book08.htm; Martial, "To Bassa," in *Epigrams*, Book 4, www.tertullian.org/fathers/martial_epigrams_book04.htm; and Martial, "On Philaenis," in *Epigrams*, Book 9, www.tertullian.org/fathers/martial_epigrams_book09.htm. Contrary to popular belief, Tyrian purple was not restricted to royalty in ancient times, only during the later Byzantine Empire.

21. Strabo, *Geography*, Vol. VII, Book XVI, sec. 23, trans Horace Leonard Jones, Loeb Classical Library (Cambridge, MA: Harvard University Press, 1954), 269, archive.org/details/in.gov.ignca.2919/page/n279/mode/2up.

22. The pH scale is logarithmic, so a solution with a pH of 8 is ten times as alkaline as one with a pH of 7.

23. Deborah Ruscillo, "Reconstructing Murex Royal Purple and Biblical Blue in the Aegean," in *Archaeomalacology: Molluscs in Former Environments of Human Behaviour*, ed. Daniella E. Bar-Yosef Mayer (Oxford: Oxbow Books, 2005), 99–106, www.academia.edu/373048/Reconstructing_Murex_Royal_Purple_and_Biblical_Blue_in_the_Aegean; Deborah Ruscillo Cosmopoulos, interview with the author, January 12, 2019.

24. Gioanventura Rosetti, *The Plictho: Instructions in the Art of the Dyers which Teaches the Dyeing of Woolen Cloths, Linens, Cottons, and Silk by the Great Art as Well as by the Common*, trans. Sidney M. Edelstein and Hector C. Borghetty (Cambridge, MA: MIT Press, 1969), 89, 91, 109–110. The translators argue that the book's strange title is probably related to the modern Italian word *plico*, meaning "envelope" or "package," and suggests a collection of instructions or important papers.

25. Cardon, *Natural Dyes*, 107–108; Zvi C. Koren (Kornblum), "Analysis of the Masada Textile Dyes," in *Masada IV. The Yigael Yadin Excavations 1963–1965. Final Reports*, ed. Joseph Aviram, Gideon Foerster, and Ehud Netzer (Jerusalem: Israel Exploration Society, 1994), 257–264.

26. Drea Leed, "Bran Water," July 2, 2003, www.elizabethancostume.net/dyes/lyteldyebook/branwater.html and "How Did They Dye Red in the Renaissance," www.elizabethancostume.net/dyes/university/renaissance_red_ingredients.pdf.

27. Koren, "Modern Chemistry of the Ancient Chemical Processing," 200–204.

28. Cardon, *Natural Dyes*, 39.

29. Cardon, *Natural Dyes*, 20–24; Charles Singer, *The Earliest Chemical Industry: An Essay in the Historical Relations of Economics and Technology Illustrated from the Alum Trade* (London: Folio Society, 1948), 114, 203–206. The quote is from Vannoccio Biringuccio in his landmark 1540 book on metalworking, *De la Pirotechnia*.

30. Rosetti, *The Plictho*, 115.

31. Mari-Tere Álvarez, "New World *Palo de Tintes* and the Renaissance Realm of Painted Cloths, Pageantry and Parade" (paper presented at From Earthly Pleasures to

Princely Glories in the Medieval and Renaissance Worlds conference, UCLA Center for Medieval and Renaissance Studies, May 17, 2013); Elena Phipps, "Global Colors: Dyes and the Dye Trade," in *Interwoven Globe: The Worldwide Textile Trade, 1500–1800*, ed. Amelia Peck (New Haven, CT: Yale University Press, 2013), 128–130.

32. Sidney M. Edelstein and Hector C. Borghetty, "Introduction," in Gioanventura Rosetti, *The Plictho*, xviii. Edelstein was a prominent industrial chemist and entrepreneur who pursued dye history as an avocation, collected many important historical works on dyeing, and gave philanthropic support to the study of chemical history and historical dyes. Anthony S. Travis, "Sidney Milton Edelstein, 1912–1994," Edelstein Center for the Analysis of Ancient Artifacts, https://edelsteincenter.wordpress.com/about/the-edelstein-center/dr-edelsteins-biography/; Drea Leed, interview with the author, January 25, 2019.

33. By the 1570s, cochineal had largely replaced kermes, but when *The Plictho* was published both reds were still in use.

34. Amy Butler Greenfield, *A Perfect Red: Empire, Espionage, and the Quest for the Color of Desire* (New York: HarperCollins, 2005), 76.

35. "The Evils of Cochineal, Tlaxcala, Mexico (1553)," in *Colonial Latin America: A Documentary History*, ed. Kenneth Mills, William B. Taylor, and Sandra Lauderdale Graham (Lanham, MD: Rowman & Littlefield, 2002), 113–116.

36. Raymond L. Lee, "Cochineal Production and Trade in New Spain to 1600," *The Americas* 4, no. 4 (April 1948): 449–473; Raymond L. Lee, "American Cochineal in European Commerce, 1526–1625," *Journal of Modern History* 23, no. 3 (September 1951): 205–224; John H. Munro, "The Medieval Scarlet and the Economics of Sartorial Splendour," in *Cloth and Clothing in Medieval Europe*, ed. N. B. Harte and K. G. Ponting (London: Heinemann Educational Books, 1983), 63–64.

37. Edward McLean Test, *Sacred Seeds: New World Plants in Early Modern Literature* (Lincoln: University of Nebraska Press, 2019), 48; Marcus Gheeraerts the Younger, *Robert Devereux, 2nd Earl of Essex*, National Portrait Gallery, www.npg.org.uk/collections/search/portrait/mw02133/Robert-Devereux-2nd-Earl-of-Essex.

38. Lynda Shaffer, "Southernization," *Journal of World History* 5 (Spring 1994): 1–21, https://roosevelt.ucsd.edu/_files/mmw/mmw12/SouthernizationArgumentAnalysis2014.pdf; Beverly Lemire and Giorgio Riello, "East & West: Textiles and Fashion in Early Modern Europe," *Journal of Social History* 41, no. 4 (Summer 2008): 887–916, http://wrap.warwick.ac.uk/190/1/WRAP_Riello_Final_Article.pdf; John Ovington, *A Voyage to Suratt: In the Year 1689* (London: Tonson, 1696), 282. Eventually, Indian cloth became such a threat to domestic textile industries that most European governments, with the notable exception of the Netherlands, restricted or banned imports. See chapter 6.

39. John J. Beer, "Eighteenth-Century Theories on the Process of Dyeing," *Isis* 51, no. 1 (March 1960): 21–30.

40. Jeanne-Marie Roland de La Platière, *Lettres de madame Roland*, 1780–1793, ed. Claude Perroud (Paris: Imprimerie Nationale, 1900), 375, https://gallica.bnf.fr/ark:/12148/bpt6k46924q/f468.item, translation by the author.

41. Société d'histoire naturelle et d'ethnographie de Colmar, *Bulletin de la Société d'histoire naturelle de Colmar: Nouvelle Série 1, 1889–1890* (Colmar: Imprimerie Decker, 1891), 282–286, https://gallica.bnf.fr/ark:/12148/bpt6k9691979j/f2.item.r=haussmann, translation by the author; Hanna Elisabeth Helvig Martinsen, *Fashionable Chemistry: The History*

of Printing Cotton in France in the Second Half of the Eighteenth and First Decades of the Nineteenth Century (PhD thesis, University of Toronto, 2015), 91–97, https://tspace.library .utoronto.ca/bitstream/1807/82430/1/Martinsen_Hanna_2015_PhD_thesis.pdf.

42. American Chemical Society, "The Chemical Revolution of Antoine-Laurent Lavoisier," June 8, 1999, www.acs.org/content/acs/en/education/whatischemistry/landmarks /lavoisier.html.

43. Martinsen, *Fashionable Chemistry*, 64.

44. Charles Coulston Gillispie, *Science and Polity in France at the End of the Old Regime* (Princeton, NJ: Princeton University Press, 1980), 409–413.

45. Claude-Louis Berthollet and Amedée B. Berthollet, *Elements of the Art of Dyeing and Bleaching*, trans. Andrew Are (London: Thomas Tegg, 1841), 284.

46. *Demorest's Family Magazine*, November 1890, 47, 49, April 1891, 381, 383, and January 1891, 185, www.google.com/books/edition/Demorest_s_Family_Magazine/dRQ7A QAAMAAJ?hl=en&gbpv=0; Diane Fagan Affleck and Karen Herbaugh, "Bright Blacks, Neon Accents: Fabrics of the 1890s," Costume Colloquium, November 2014.

47. John W. Servos, "The Industrialization of Chemistry," *Science* 264, no. 5161 (May 13, 1994): 993–994.

48. Catherine M. Jackson, "Synthetical Experiments and Alkaloid Analogues: Liebig, Hofmann, and the Origins of Organic Synthesis," *Historical Studies in the Natural Sciences* 44, no. 4 (September 2014): 319–363; Augustus William Hofmann, "A Chemical Investigation of the Organic Bases contained in Coal-Gas," *London, Edinburgh, and Dublin Philosophical Magazine and Journal of Science*, February 1884, 115–127; W. H. Perkin, "The Origin of the Coal-Tar Colour Industry, and the Contributions of Hofmann and His Pupils," *Journal of the Chemical Society*, 1896, 596–637.

49. Sir F. A. Abel, "The History of the Royal College of Chemistry and Reminiscences of Hofmann's Professorship," *Journal of the Chemical Society*, 1896, 580–596.

50. Anthony S. Travis, "Science's Powerful Companion: A. W. Hofmann's Investigation of Aniline Red and Its Derivatives," *British Journal for the History of Science* 25, no. 1 (March 1992): 27–44; Edward J. Hallock, "Sketch of August Wilhelm Hofmann," *Popular Science Monthly*, April 1884, 831–835; Lord Playfair, "Personal Reminiscences of Hofmann and of the Conditions which Led to the Establishment of the Royal College of Chemistry and His Appointment as Its Professor," *Journal of the Chemical Society*, 1896, 575–579; Anthony S. Travis, *The Rainbow Makers: The Origins of the Synthetic Dyestuffs Industry in Western Europe* (Bethlehem, NY: Lehigh University Press, 1993), 31–81, 220–227.

51. Simon Garfield, *Mauve: How One Man Invented a Colour That Changed the World* (London: Faber & Faber, 2000), 69.

52. Travis, *The Rainbow Makers*, 31–81, 220–227; Perkin, "The Origin of the Coal-Tar Colour Industry."

53. Robert Chenciner, *Madder Red: A History of Luxury and Trade* (London: Routledge Curzon, 2000), Kindle locations 5323–5325; J. E. O'Conor, *Review of the Trade of India, 1900–1901* (Calcutta: Office of the Superintendent of Government Printing, 1901), 28–29; Asiaticus, "The Rise and Fall of the Indigo Industry in India," *Economic Journal*, June 1912, 237–247.

54. Somaiya Kala Vidya is primarily a school training accomplished artisans in better design and marketing practices, but it also runs workshops for interested amateurs, such as the one I attended February 27–March 10, 2019, www.somaiya-kalavidya.org/about.html.

55. There are technically four separate companies: Swisstex California, the original dye house; Swisstex Direct, a fabric company that buys yarns and contracts out knitting; Swisstex El Salvador, a dye house in that country; and Unique, a fabric manufacturer in El Salvador. Dyeing is more important in Los Angeles, while fabric predominates in El Salvador, close to where garments are assembled. All are owned equally by the same four partners. Dartley is the president of Swisstex Direct.

56. Badri Chatterjee, "Why Are Dogs Turning Blue in This Mumbai Suburb? Kasadi River May Hold Answers," *Hindustan Times*, August 11, 2017, www.hindustantimes.com/mumbai-news/industrial-waste-in-navi-mumbai-s-kasadi-river-is-turning-dogs-blue/story-FcG0fUpioHGWUY1zv98HuN.html; Badri Chatterjee, "Mumbai's Blue Dogs: Pollution Board Shuts Down Dye Industry After HT Report," *Hindustan Times*, August 20, 2017, www.hindustantimes.com/mumbai-news/mumbai-s-blue-dogs-pollution-board-shuts-down-dye-industry-after-ht-report/story-uhgaiSeIk7UbxV93WLniaN.html.

57. Keith Dartley, interviews with the author, September 16, 2019, and September 26, 2019, and email to the author, September 27, 2019; Swisstex California, "Environment," www.swisstex-ca.com/Swisstex_Ca/Environment.html. Swisstex is certified by Bluesign, an environmental standard-setting and monitoring company based in Switzerland: www.bluesign.com/en.

Chapter Five: Traders

1. Cécile Michel, *Correspondance des marchands de Kaniš au début du IIe millénaire avant J.-C.* (Paris: Les Éditions du Cerf, 2001), 427–431 (translation from French by the author); Cécile Michel, "The Old Assyrian Trade in the Light of Recent Kültepe Archives," *Journal of the Canadian Society for Mesopotamian Studies*, 2008, 71–82, https://halshs.archives-ouvertes.fr/halshs-00642827/document; Cécile Michel, "Assyrian Women's Contribution to International Trade with Anatolia," *Carnet de REFEMA*, November 12, 2013, https://refema.hypotheses.org/850; Cécile Michel, "Economic and Social Aspects of the Old Assyrian Loan Contract," in *L'economia dell'antica Mesopotamia (III-I millennio a.C.) Per un dialogo interdisciplinare*, ed. Franco D'Agostino (Rome: Edizioni Nuova Cultura, 2013), 41–56, https://halshs.archives-ouvertes.fr/halshs-01426527/document; Mogens Trolle Larsen, *Ancient Kanesh: A Merchant Colony in Bronze Age Anatolia* (Cambridge: University of Cambridge Press, 2015), 1–3, 112, 152–158, 174, 196–201; Klaas R. Veenhof, "'Modern' Features in Old Assyrian Trade," *Journal of the Economic and Social History of the Orient* 40, no. 4 (January 1997): 336–366.

2. On social technology, see Richard R. Nelson, "Physical and Social Technologies, and Their Evolution" (LEM Working Paper Series, Scuola Superiore Sant'Anna, Laboratory of Economics and Management [LEM], Pisa, Italy, June 2003), http://hdl.handle.net/10419/89537.

3. Larsen, *Ancient Kanesh*, 54–57.

4. Larsen, *Ancient Kanesh*, 181–182.

5. Jessica L. Goldberg, *Trade and Institutions in the Medieval Mediterranean: The Geniza Merchants and Their Business World* (Cambridge: Cambridge University Press, 2012), 65.

6. The Sogdians, a Central Asian people whose major cities were Samarkand and Bukhara in what is now Uzbekistan, were important traders between China and Iran.

7. Valerie Hansen and Xinjiang Rong, "How the Residents of Turfan Used Textiles as Money, 273–796 CE," *Journal of the Royal Asiatic Society* 23, no. 2 (April 2013): 281–305,

https://history.yale.edu/sites/default/files/files/VALERIE%20HANSEN%20and%20XIN
JIANG%20RONG.pdf.

8. Chang Xu and Helen Wang (trans.), "Managing a Multicurrency System in Tang
China: The View from the Centre," *Journal of the Royal Asiatic Society* 23, no. 2 (April 2013):
242.

9. Such a story was called a *þáttr*, the Old Norse word for "strand," as in a piece of yarn.

10. William Ian Miller, *Audun and the Polar Bear: Luck, Law, and Largesse in a Medieval
Tale of Risky Business* (Leiden: Brill, 2008), 7, 22–25.

11. As a unit of account, the law specified that a standard piece of *vaðmál* was equivalent
to an ounce of silver.

12. Michèle Hayeur Smith, "*Vaðmál* and Cloth Currency in Viking and Medieval Iceland," in *Silver, Butter, Cloth: Monetary and Social Economies in the Viking Age*, ed. Jane
Kershaw and Gareth Williams (Oxford: Oxford University Press, 2019), 251–277; Michèle
Hayeur Smith, "Thorir's Bargain: Gender, *Vaðmál* and the Law," *World Archaeology* 45,
no. 5 (2013): 730–746, https://doi.org/10.1080/00438243.2013.860272; Michèle Hayeur
Smith, "Weaving Wealth: Cloth and Trade in Viking Age and Medieval Iceland," in *Textiles and the Medieval Economy: Production, Trade, and Consumption of Textiles, 8th–16th
Centuries*, ed. Angela Ling Huang and Carsten Jahnke (Oxford: Oxbow Books, 2014),
23–40, www.researchgate.net/publication/272818539_Weaving_Wealth_Cloth_and
_Trade_in_Viking_Age_and_Medieval_Iceland. Although often used as a notional unit
of account, sometimes called "ghost money," actual silver was exchanged much less often
than cloth. For a simple overview of the basic characteristics of money, see Federal Reserve
Bank of St. Louis, "Functions of Money," Economic Lowdown Podcast Series, Episode
9, www.stlouisfed.org/education/economic-lowdown-podcast-series/episode-9-functions-of
-money.

13. Marion Johnson, "Cloth as Money: The Cloth Strip Currencies of Africa," *Textile
History* 11, no. 1 (1980): 193–202.

14. Peter Spufford, *Power and Profit: The Merchant in Medieval Europe* (London: Thames
& Hudson, 2002), 134–136, 143–152.

15. Alessandra Macinghi degli Strozzi, *Lettere di una Gentildonna Fiorentina del Secolo
XV ai Figliuoli Esuli*, ed. Cesare Guasti (Firenze: G. C. Sansone, 1877), 27–30. (Translation
by the author.)

16. Spufford, *Power and Profit*, 25–29.

17. Jong Kuk Nam, "The *Scarsella* between the Mediterranean and the Atlantic in the
1400s," *Mediterranean Review*, June 2016, 53–75.

18. Telesforo Bini, "Lettere mercantili del 1375 di Venezia a Giusfredo Cenami setai-
olo," appendix to *Su I lucchesi a Venezia: Memorie dei Secoli XII e XIV*, Part 2, in *Atti dell'Ac-
cademia Lucchese di Scienze, Lettere ed Arti* (Lucca, Italy: Tipografia di Giuseppe Giusti,
1857), 150–155, www.google.com/books/edition/_/OLwAAAAAYAAJ?hl=en.

19. Spufford, *Power and Profit*, 28–29.

20. Warren Van Egmond, *The Commercial Revolution and the Beginnings of Western
Mathematics in Renaissance Florence, 1300–1500* (unpublished dissertation, History and
Philosophy of Science, Indiana University, 1976), 74–75, 106. Much of the following draws
from Van Egmond's research. I give page numbers only for quotations and a few specific
facts.

21. Van Egmond, *The Commercial Revolution*, 14, 172, 186–187, 196–197, 251.

22. L. E. Sigler, *Fibonacci's Liber Abaci: A Translation into Modern English of Leonardo Pisano's Book of Calculation* (New York: Springer-Verlag, 2002), 4, 15–16.

23. Paul F. Grendler, *Schooling in Renaissance Italy: Literacy and Learning, 1300–1600* (Baltimore: Johns Hopkins University Press, 1989), 77, 306–329; Margaret Spufford, "Literacy, Trade, and Religion in the Commercial Centers of Europe," in *A Miracle Mirrored: The Dutch Republic in European Perspective*, ed. Karel A. Davids and Jan Lucassen (Cambridge: Cambridge University Press, 1995), 229–283. Paul F. Grendler, "What Piero Learned in School: Fifteenth-Century Vernacular Education," *Studies in the History of Art* (Symposium Papers XXVIII: Piero della Francesca and His Legacy, 1995), 160–174; Frank J. Swetz, *Capitalism and Arithmetic: The New Math of the 15th Century, Including the Full Text of the Treviso Arithmetic of 1478*, trans. David Eugene Smith (La Salle, IL: Open Court, 1987).

24. Edwin S. Hunt and James Murray, *A History of Business in Medieval Europe, 1200–1550* (Cambridge: Cambridge University Press, 1999), 57–63.

25. Van Egmond, *The Commercial Revolution*, 17–18, 173.

26. Today the word *factor* has a technical meaning in the apparel business, referring to an entity that provides credit based on a manufacturer's current invoices. For most of textile history, however, it simply meant an agent or middleman.

27. James Stevens Rogers, *The Early History of the Law of Bills and Notes: A Study of the Origins of Anglo-American Commercial Law* (Cambridge: Cambridge University Press, 1995), 104–106.

28. Hunt and Murray, *A History of Business in Medieval Europe*, 64.

29. Francesca Trivellato, *The Promise and Peril of Credit: What a Forgotten Legend About Jews and Finance Tells Us About the Making of European Commercial Society* (Princeton, NJ: Princeton University Press, 2019), 2. Despite the instruments' Italian origins, the legend arose that Jews had invented bills of exchange as a way of spiriting their wealth out of Spain when they were expelled in 1492. Trivellato's book explores the origin and persistence of that legend.

30. Spufford, *Power and Profit*, 37. As they evolved and eventually became negotiable, bills of exchange came increasingly close to what economists would count as part of the money supply.

31. Meir Kohn, "Bills of Exchange and the Money Market to 1600" (Department of Economics Working Paper No. 99-04, Dartmouth College, Hanover, NH, February 1999), 21, cpb-us-e1.wpmucdn.com/sites.dartmouth.edu/dist/6/1163/files/2017/03/99-04.pdf; Peter Spufford, *Handbook of Medieval Exchange* (London: Royal Historical Society, 1986), xxxvii.

32. Spufford, *Handbook of Medieval Exchange*, 316, 321.

33. Kohn, "Bills of Exchange and the Money Market," 3, 7–9; Trivellato, *The Promise and Peril of Credit*, 29–30. See also Raymond de Roover, "What Is Dry Exchange: A Contribution to the Study of English Mercantilism," in *Business, Banking, and Economic Thought in Late Medieval and Early Modern Europe: Selected Studies of Raymond de Roover*, ed. Julius Kirshner (Chicago: University of Chicago Press, 1974), 183–199.

34. Iris Origo, *The Merchant of Prato: Daily Life in a Medieval City* (New York: Penguin, 1963), 146–149.

35. Hunt and Murray, *A History of Business in Medieval Europe*, 222–225; K. S. Mathew, *Indo-Portuguese Trade and the Fuggers of Germany: Sixteenth Century* (New Delhi: Manohar, 1997), 101–147.

36. Kohn, "Bills of Exchange and the Money Market," 28.

37. Alfred Wadsworth and Julia de Lacy Mann, *The Cotton Trade and Industrial Lancashire 1600–1780* (Manchester, UK: Manchester University Press, 1931), 91–95.

38. Wadsworth and Mann, *The Cotton Trade and Industrial Lancashire*, 91–95; T. S. Ashton, "The Bill of Exchange and Private Banks in Lancashire, 1790–1830," *Economic History Review* a15, nos. 1–2 (1945): 27.

39. Trivellato, *The Promise and Peril of Credit*, 13–14.

40. John Graham, "History of Printworks in the Manchester District from 1760 to 1846," quoted in J. K. Horsefield, "Gibson and Johnson: A Forgotten Cause Célèbre," *Economica*, August 1943, 233–237.

41. Trivellato, *The Promise and Peril of Credit*, 32–34; Kohn, "Bills of Exchange and the Money Market," 24–28; Lewis Loyd testimony, May 4, 1826, in House of Commons, *Report from the Select Committee on Promissory Notes in Scotland and Ireland* (London: Great Britain Parliament, May 26, 1826), 186.

42. Alexander Blair testimony, March 21, 1826, in House of Commons, *Report from the Select Committee on Promissory Notes in Scotland and Ireland* (London: Great Britain Parliament, May 26, 1826), 41; Lloyds Banking Group, "British Linen Bank (1746–1999)," www.lloydsbankinggroup.com/Our-Group/our-heritage/our-history2/bank-of-scotland /british-linen-bank/.

43. Carl J. Griffin, *Protest, Politics and Work in Rural England, 1700–1850* (London: Palgrave Macmillan, 2013), 24; Adrian Randall, *Riotous Assemblies: Popular Protest in Hanoverian England* (Oxford: Oxford University Press, 2006), 141–143; David Rollison, *The Local Origins of Modern Society: Gloucestershire 1500–1800* (London: Routledge, 2005), 226–227.

44. The term *woolen* refers to a heavier wool cloth that was *fulled*, a process using moisture and friction to create a felted surface. Once fulled, woolens were sheared to create a smooth surface. Woolens used soft yarn spun from carded, short-stapled wool. *Worsted* refers to lighter wool cloth, usually unfulled, woven with tightly spun thread; before spinning, the wool was combed rather than carded. Carding fluffs up fibers, whereas combing aligns them in the same direction.

45. "An Essay on Riots; Their Causes and Cure," *Gentleman's Magazine*, January 1739, 7–10. See also, "A Letter on the Woollen Manufacturer," *Gentleman's Magazine*, February 1739, 84–86; A Manufacturer in Wiltshire, "Remarks on the Essay on Riots," *Gentleman's Magazine*, March 1739, 123–126; Trowbridge, "Conclusion," *Gentleman's Magazine*, 126; "Case between the Clothiers and Weavers," *Gentleman's Magazine*, April 1739, 205–206; "The Late Improvements of Our Trade, Navigation, and Manufactures," *Gentleman's Magazine*, September 1739, 478–480.

46. Trowbridge, untitled essay, *Gentleman's Magazine*, February 1739, 89–90. Trowbridge, "Conclusion," *Gentleman's Magazine*, 126.

47. Ray Bert Westerfield, *The Middleman in English Business* (New Haven, CT: Yale University Press, 1914), 296, archive.org/details/middlemeninengli00west. Although it did not explicitly restrict their numbers, the law recognizing the factor's role and establishing

a register might have been used to do so, which would have given incumbents greater economic power.

48. Luca Molà, *The Silk Industry of Renaissance Venice* (Baltimore: Johns Hopkins University Press, 2000), 365n11.

49. Conrad Gill, "Blackwell Hall Factors, 1795–1799," *Economic History Review*, August 1954, 268–281; Westerfield, *The Middleman in English Business*, 273–304.

50. Trowbridge, "Conclusion," *Gentleman's Magazine*, 126.

51. All quotations are transcribed from *The Lehman Trilogy* as performed at Park Avenue Armory, New York, April 4, 2019.

52. Harold D. Woodman, "The Decline of Cotton Factorage After the Civil War," *American Historical Review* 71, no. 4 (July 1966): 1219–1236; Harold D. Woodman, *King Cotton and His Retainers: Financing and Marketing the Cotton Crop of the South, 1800–1925* (Lexington: University of Kentucky Press, 1968). Woodman dates cotton factors in the South back to at least 1800.

53. Italian Playwrights Project, "Stefano Massini's SOMETHING ABOUT THE LEHMANS," YouTube video, 1:34:04, December 5, 2016, www.youtube.com/watch?time_continue=112&v=gETKm6El85o.

54. Ben Brantley, "'The Lehman Trilogy' Is a Transfixing Epic of Riches and Ruin," *New York Times*, July 13, 2018, C5, www.nytimes.com/2018/07/13/theater/lehman-trilogy-review-national-theater-london.html; Richard Cohen, "The Hole at the Heart of 'The Lehman Trilogy,'" *Washington Post*, April 8, 2019, www.washingtonpost.com/opinions/the-hole-at-the-heart-of-the-lehman-trilogy/2019/04/08/51f6ed8c-5a3e-11e9-842d-7d3ed7eb3957_story.html?utm_term=.257ef2349d55; Jonathan Mandell, "The Lehman Trilogy Review: 164 Years of One Capitalist Family Minus the Dark Parts," *New York Theater*, April 7, 2019, https://newyorktheater.me/2019/04/07/the-lehman-trilogy-review-164-years-of-one-capitalist-family-minus-the-dark-parts/; Nicole Gelinas, "The Lehman Elegy," *City Journal*, April 12, 2019, www.city-journal.org/the-lehman-trilogy.

Chapter Six: Consumers

1. Angela Yu-Yun Sheng, "Textile Use, Technology, and Change in Rural Textile Production in Song, China (960–1279)" (unpublished dissertation, University of Pennsylvania, 1990), 53, 68–113.

2. Roslyn Lee Hammers, *Pictures of Tilling and Weaving: Art, Labor, and Technology in Song and Yuan China* (Hong Kong: Hong Kong University Press, 2011), 1–7, 87–98, 210, 211. Translations are by Hammers and published with her kind permission.

3. The Song dynasty lasted from 960 to 1279 CE. It is divided into the Northern Song, which ended in 1127 when the Jurchen Jin dynasty conquered northern China, including the capital in what is now Kaifeng. The Southern Song governed China south of the Yangtze from a new capital in what is now Hangzhou. The movement of officials and much of the population to the south permanently altered the economic geography of China.

4. William Guanglin Liu, *The Chinese Market Economy 1000–1500* (Albany: State University of New York Press, 2015), 273–275; Richard von Glahn, *The Economic History of China: From Antiquity to the Nineteenth Century* (Cambridge: Cambridge University Press, 2016), 462.

5. Liu, *The Chinese Market Economy*, 273–278; Sheng, "Textile Use, Technology, and Change," 174.

6. Thomas T. Allsen, *Commodity and Exchange in the Mongol Empire: A Cultural History of Islamic Textiles* (Cambridge: Cambridge University Press, 1997), 28; Sheila S. Blair, "East Meets West Under the Mongols," *Silk Road* 3, no. 2 (December 2005): 27–33, www .silkroadfoundation.org/newsletter/vol3num2/6_blair.php.

7. The Tartars were one of Genghis Khan's first conquests and were absorbed into the new Mongol identity he referred to as "the People of the Felt Walls." Jack Weatherford, *Genghis Khan and the Making of the Modern World* (New York: Crown, 2004), 53–54.

8. Joyce Denney, "Textiles in the Mongol and Yuan Periods," and James C. Y. Watt, "Introduction," in James C. Y. Watt, *The World of Khubilai Khan: Chinese Art in the Yuan Dynasty* (New York: Metropolitan Museum of Art, 2010), 243–267, 7–10.

9. Peter Jackson, *The Mongols and the Islamic World* (New Haven, CT: Yale University Press, 2017), 225; Allsen, *Commodity and Exchange in the Mongol Empire*, 38–45, 101; Denney, "Textiles in the Mongol and Yuan Periods."

10. Helen Persson, "Chinese Silks in Mamluk Egypt," in *Global Textile Encounters*, ed. Marie-Louise Nosch, Zhao Feng, and Lotika Varadarajan (Oxford: Oxbow Books, 2014), 118.

11. James C. Y. Watt and Anne E. Wardwell, *When Silk Was Gold: Central Asian and Chinese Textiles* (New York: Metropolitan Museum of Art, 1997), 132.

12. Allsen, *Commodity and Exchange in the Mongol Empire*, 29. Genghis Khan urged his commanders to train their sons so they would be as knowledgeable in the arts of war as merchants were about their goods.

13. Yuan Zujie, "Dressing the State, Dressing the Society: Ritual, Morality, and Conspicuous Consumption in Ming Dynasty China" (unpublished dissertation, University of Minnesota, 2002), 51.

14. Craig Clunas, *Superfluous Things: Material Culture and Social Status in Early Modern China* (Urbana: University of Illinois Press, 1991), 150; Zujie, "Dressing the State, Dressing the Society," 93.

15. The primary revision occurred in 1528, establishing rules for the off-duty attire of officials.

16. BuYun Chen, "Wearing the Hat of Loyalty: Imperial Power and Dress Reform in Ming Dynasty China," in *The Right to Dress: Sumptuary Laws in a Global Perspective, c. 1200–1800*, ed. Giorgio Riello and Ulinka Rublack (Cambridge: Cambridge University Press, 2019), 418.

17. Zujie, "Dressing the State, Dressing the Society," 94–96, 189–191.

18. Ulinka Rublack, "The Right to Dress: Sartorial Politics in Germany, c. 1300–1750," in *The Right to Dress*, 45; Chen, "Wearing the Hat of Loyalty," 430–431.

19. Liza Crihfield Darby, *Kimono: Fashioning Culture* (Seattle: University of Washington Press, 2001), 52–54; Katsuya Hirano, "Regulating Excess: The Cultural Politics of Consumption in Tokugawa Japan," in *The Right to Dress*, 435–460; Howard Hibbett, *The Floating World in Japanese Fiction* (Boston: Tuttle Publishing, [1959] 2001).

20. Catherine Kovesi, "Defending the Right to Dress: Two Sumptuary Law Protests in Sixteenth-Century Milan," in *The Right to Dress*, 186; Luca Molà and Giorgio Riello, "Against the Law: Sumptuary Prosecutions in Sixteenth- and Seventeenth-Century

Padova," in *The Right to Dress*, 216; Maria Giuseppina Muzzarelli, "Sumptuary Laws in Italy: Financial Resource and Instrument of Rule," in *The Right to Dress*, 171, 176; Alan Hunt, *Governance of the Consuming Passions: A History of Sumptuary Law* (New York: St. Martin's Press, 1996), 73; Ronald E. Rainey, "Sumptuary Legislation in Renaissance Florence" (unpublished diss., Columbia University, 1985), 62.

21. Rainey, "Sumptuary Legislation in Renaissance Florence," 54, 468–470, 198.

22. Rainey, "Sumptuary Legislation in Renaissance Florence," 52–53, 72, 98, 147, 442–443. The word *sciamito* may refer specifically to *samite*, a reversible brocade often including gold or silver threads, but Rainey finds that it was also used in a more general sense.

23. Carole Collier Frick, *Dressing Renaissance Florence: Families, Fortunes, and Fine Clothing* (Baltimore: Johns Hopkins University Press, 2005), Kindle edition.

24. Rainey, "Sumptuary Legislation in Renaissance Florence," 231–234; Franco Sacchetti, *Tales from Sacchetti*, trans. by Mary G. Steegman (London: J. M. Dent, 1908), 117–119; Franco Sacchetti, *Delle Novelle di Franco Sacchetti* (Florence: n.p., 1724), 227. The original phrase is "Ciò che vuole dunna [*sic*], vuol signò; e ciò vuol signò, Tirli in Birli."

25. Muzzarelli, "Sumptuary Laws in Italy," 175, 185.

26. Rainey, "Sumptuary Legislation in Renaissance Florence," 200–205, 217; William Caferro, "Florentine Wages at the Time of the Black Death" (unpublished ms., Vanderbilt University), https://economics.yale.edu/sites/default/files/florence_wages-caferro.pdf.

27. Kovesi, "Defending the Right to Dress," 199–200.

28. Felicia Gottmann, *Global Trade, Smuggling, and the Making of Economic Liberalism: Asian Textiles in France 1680–1760* (Basingstoke, UK: Palgrave Macmillan, 2016), 91. A version of this section originally appeared as Virginia Postrel, "Before Drug Prohibition, There Was the War on Calico," *Reason*, July 2018, 14–15, https://reason.com/2018/06/25/before-drug-prohibition-there/.

29. Michael Kwass, *Contraband: Louis Mandrin and the Making of a Global Underground* (Cambridge, MA: Harvard University Press, 2014), 218–220; Gillian Crosby, *First Impressions: The Prohibition on Printed Calicoes in France, 1686–1759* (unpublished dissertation, Nottingham Trent University, 2015), 143–144.

30. Kwass, *Contraband*, 56.

31. For an overview of the historiography on the British Calico Acts, including links to relevant literature, see "The Calico Acts: Was British Cotton Made Possible by Infant Industry Protection from Indian Competition?" Pseudoerasmus, January 5, 2017, https://pseudoerasmus.com/2017/01/05/ca/.

32. Giorgio Riello, *Cotton: The Fabric That Made the Modern World* (Cambridge: Cambridge University Press, 2013), 100; Kwass, *Contraband*, 33.

33. Gottmann, *Global Trade, Smuggling*, 7; Kwass, *Contraband*, 37–39.

34. Gottmann, *Global Trade, Smuggling*, 41.

35. Gottmann, *Global Trade, Smuggling*, 153.

36. Kwass, *Contraband*, 294.

37. Julie Gibbons, "The History of Surface Design: Toile de Jouy," Pattern Observer, https://patternobserver.com/2014/09/23/history-surface-design-toile-de-jouy/.

38. George Metcalf, "A Microcosm of Why Africans Sold Slaves: Akan Consumption Patterns in the 1770s," *Journal of African History* 28, no. 3 (November 1987): 377–394. The popularity of textiles is confirmed in data compiled in Stanley B. Alpern, "What Africans

Got for Their Slaves: A Master List of European Trade Goods," *History in Africa* 22 (January 1995): 5–43.

39. In this period, most West African captives were bound for the sugar plantations of the West Indies.

40. Chambon, *Le commerce de l'Amérique par Marseille*, quoted and translated in Michael Kwass, *Contraband*, 20. Original available at https://gallica.bnf.fr/ark:/12148 /bpt6k1041911g/f417.item.zoom. Venice Lamb, *West African Weaving* (London: Duckworth, 1975), 104.

41. Colleen E. Kriger, "'Guinea Cloth': Production and Consumption of Cotton Textiles in West Africa before and during the Atlantic Slave Trade," in *The Spinning World: A Global History of Cotton Textiles, 1200–1850*, ed. Giorgio Riello and Prasannan Parthasarathi (Oxford: Oxford University Press, 2009), 105–126; Colleen E. Kriger, *Cloth in West African History* (Lanham, MD: Altamira Press, 2006), 35–36.

42. Suzanne Gott and Kristyne S. Loughran, "Introducing African-Print Fashion," in *African-Print Fashion Now! A Story of Taste, Globalization, and Style*, ed. Suzanne Gott, Kristyne S. Loughran, Betsy D. Smith, and Leslie W. Rabine (Los Angeles: Fowler Museum UCLA, 2017), 22–49; Helen Elanda, "Dutch Wax Classics: The Designs Introduced by Ebenezer Brown Fleming circa 1890–1912 and Their Legacy," in *African-Print Fashion Now!*, 52–61; Alisa LaGamma, "The Poetics of Cloth," in *The Essential Art of African Textiles: Design Without End*, ed. Alisa LaGamma and Christine Giuntini (New Haven, CT: Yale University Press, 2008), 9–23, www.metmuseum.org/art/metpublications /the_essential_art_of_african_textiles_design_without_end.

43. Kathleen Bickford Berzock, "African Prints/African Ownership: On Naming, Value, and Classics," in *African-Print Fashion Now!*, 71–79. (Berzock is the art historian quoted.) Susan Domowitz, "Wearing Proverbs: Anyi Names for Printed Factory Cloth," *African Arts*, July 1992, 82–87, 104; Paulette Young, "Ghanaian Woman and Dutch Wax Prints: The Counter-appropriation of the Foreign and the Local Creating a New Visual Voice of Creative Expression," *Journal of Asian and African Studies* 51, no. 3 (January 10, 2016), https://doi.org/10.1177/0021909615623811. (Young is the curator quoted.) Michelle Gilbert, "Names, Cloth and Identity: A Case from West Africa," in *Media and Identity in Africa*, ed. John Middleton and Kimani Njogu (Bloomington: Indiana University Press, 2010), 226–244.

44. Tunde M. Akinwumi, "The 'African Print' Hoax: Machine Produced Textiles Jeopardize African Print Authenticity," *Journal of Pan African Studies* 2, no. 5 (July 2008): 179–192; Victoria L. Rovine, "Cloth, Dress, and Drama," in *African-Print Fashion Now!*, 274–277.

45. Although described as kente by Colleen Kriger, this cloth may be better seen as a precursor to it. Malika Kraamer, "Ghanaian Interweaving in the Nineteenth Century: A New Perspective on Ewe and Asante Textile History," *African Arts*, Winter 2006, 36–53, 93–95.

46. Depending on how the weft is inserted, the patterns may not appear as vertical and horizontal. That is simply the most common arrangement. "Just to confuse matters, it is, of course, possible to weave warp-wise stripes in weft-faced areas by using alternating weft elements in two colours. Those in the one colour will then all lie over one warp unit and under

the next, and vice versa for weft elements in the other colour," observe textile scholars John Picton and John Mack, *African Textiles* (New York: Harper & Row, 1989), 117.

47. Malika Kraamer, "Challenged Pasts and the Museum: The Case of Ghanaian *Kente*," in *The Thing about Museums: Objects and Experience, Representation and Contestation*, ed. Sandra Dudley, Amy Jane Barnes, Jennifer Binnie, Julia Petrov, Jennifer Walklate (Abingdon, UK: Routledge, 2011), 282–296.

48. Lamb, *West African Weaving*, 141.

49. Lamb, *West African Weaving*, 22; Doran H. Ross, "Introduction: Fine Weaves and Tangled Webs" and "Kente and Its Image Outside Ghana," in *Wrapped in Pride: Ghanaian Kente and African American Identity*, ed. Doran H. Ross (Los Angeles: UCLA Fowler Museum of Cultural History, 1998), 21, 160–176; James Padilioni Jr., "The History and Significance of Kente Cloth in the Black Diaspora," Black Perspectives, May 22, 2017, www.aaihs.org/the-history-and-significance-of-kente-cloth-in-the-black-diaspora//; Betsy D. Quick, "Pride and Dignity: African American Perspective on Kente," in *Wrapped in Pride*, 202–268. Kente can be seen as a manifestation of glamour. See Virginia Postrel, *The Power of Glamour: Longing and the Art of Visual Persuasion* (New York: Simon & Schuster, 2013).

50. Anita M. Samuels, "African Textiles: Making the Transition from Cultural Statement to Macy's," *New York Times*, July 26, 1992, sec. 3, 10, www.nytimes.com/1992/07/26/business/all-about-african-textiles-making-transition-cultural-statement-macy-s.html. Perhaps the importer or the reporter confused kente with wax prints, which are printed on both sides of the fabric.

51. Ross, *Wrapped in Pride*, 273–289.

52. Kwesi Yankah, "Around the World in Kente Cloth," *Uhuru*, May 1990, 15–17, quoted in Ross, *Wrapped in Pride*, 276; John Picton, "Tradition, Technology, and Lurex: Some Comments on Textile History and Design in West Africa," in *History, Design, and Craft in West African Strip-Woven Cloth: Papers Presented at a Symposium Organized by the National Museum of African Art, Smithsonian Institution, February 18–19, 1988* (Washington, DC: Smithsonian Institution, 1992), 46. For kente yoga pants, see www.etsy.com/market/kente_leggings. For a fuller discussion of authenticity, see Virginia Postrel, *The Substance of Style: How the Rise of Aesthetic Value Is Remaking Culture, Commerce, and Consciousness* (New York: HarperCollins, 2003), 95–117.

53. For an American textile example, see Virginia Postrel, "Making History Modern," *Reason*, December 2017, 10–11, https://vpostrel.com/articles/making-history-modern. For a Mexican example, see Virginia Postrel, "How Ponchos Got More Authentic After Commerce Came to Chiapas," *Reason*, April 2018, 10–11, https://vpostrel.com/articles/how-ponchos-got-more-authentic-after-commerce-came-to-chiapas.

54. Raymond Senuk, interview with the author, August 31, 2018, and email August 2, 2019; Lisa Fitzpatrick, interview with the author, August 24, 2018; Barbara Knoke de Arathoon and Rosario Miralbés de Polanco, *Huipiles Mayas de Guatemala/Maya Huipiles of Guatemala* (Guatemala City: Museo Ixchel del Traje Indigene, 2011); Raymond E. Senuk, *Maya Traje: A Tradition in Transition* (Princeton, NJ: Friends of the Ixchel Museum, 2019); Rosario Miralbés de Polanco, *The Magic and Mystery of Jaspe: Knots Revealing Designs* (Guatemala City: Museo Ixchel del Traje Indigena, 2005). On Instagram, see www.instagram.com/explore/tags/chicasdecorte/.

55. Chris Anderson, *The Long Tail: Why the Future of Business Is Selling Less of More* (New York: Hachette Books, 2008), 52.

56. Gart Davis, interview with the author, May 11, 2016, and email to the author, August 2, 2019; Alex Craig email to the author, September 23, 2019; Jonna Hayden, Facebook messages with the author, May 10, 2016, and August 3, 2019.

Chapter Seven: Innovators

1. Sharon Bertsch McGrayne, *Prometheans in the Lab: Chemistry and the Making of the Modern World* (New York: McGraw-Hill, 2001), 114. Some of the following material appeared in Virginia Postrel, "The iPhone of 1939 Helped Liberate Europe. And Women," *Bloomberg Opinion*, October 25, 2019, www.bloomberg.com/opinion/articles/2019-10-25 /nylon-history-how-stockings-helped-liberate-women.

2. Yasu Furukawa, *Inventing Polymer Science: Staudinger, Carothers, and the Emergence of Macromolecular Chemistry* (Philadelphia: University of Pennsylvania Press, 1998), 103–111; Joel Mokyr, *The Gifts of Athena: Historical Origins of the Knowledge Economy* (Princeton, NJ: Princeton University Press, 2002), 28–77.

3. Herman F. Mark, "The Early Days of Polymer Science," in *Contemporary Topics in Polymer Science*, Vol. 5, ed. E.J. Vandenberg, Proceedings of the Eleventh Biennial Polymer Symposium of the Division of Polymer Chemistry on High Performance Polymers, November 20–24, 1982 (New York: Plenum Press, 1984), 10–11.

4. McGrayne, *Prometheans in the Lab*, 120–128; Matthew E. Hermes, *Enough for One Lifetime: Wallace Carothers, Inventor of Nylon* (Washington, DC: American Chemical Society and Chemical Heritage Foundation, 1996), 115.

5. "Chemists Produce Synthetic 'Silk,'" *New York Times*, September 2, 1931, 23.

6. Hermes, *Enough for One Lifetime*, 183.

7. McGrayne, *Prometheans in the Lab*, 139–142; Hermes, *Enough for One Lifetime*, 185–189.

8. "The New Dr. West's Miracle Tuft" ad, *Saturday Evening Post*, October 29, 1938, 44–45, https://archive.org/details/the-saturday-evening-post-1938-10-29/page/n43; "DuPont Discloses New Yarn Details," *New York Times*, October 28, 1938, 38; "Du Pont Calls Fair American Symbol," *New York Times*, April 25, 1939, 2; "First Offering of Nylon Hosiery Sold Out," *New York Times*, October 25, 1939, 38; "Stine Says Nylon Claims Tend to Overoptimism," *New York Times*, January 13, 1940, 18.

9. Kimbra Cutlip, "How 75 Years Ago Nylon Stockings Changed the World," *Smithsonian*, May 11, 2015, www.smithsonianmag.com/smithsonian-institution/how -75-years-ago-nylon-stockings-changed-world-180955219/.

10. David Brunnschweiler, "Rex Whinfield and James Dickson at the Broad Oak Print Works," in *Polyester: 50 Years of Achievement*, ed. David Brunnschweiler and John Hearle (Manchester, UK: Textile Institute, 1993), 34–37; J. R. Whinfield, "The Development of Terylene," *Textile Research Journal*, May 1953, 289–293, https://doi .org/10.1177/004051755302300503; J. R. Whinfield, "Textiles and the Inventive Spirit" (Emsley Lecture), in *Journal of the Textile Institute Proceedings*, October 1955, 5–11; IHS Markit, "Polyester Fibers," *Chemical Economics Handbook*, June 2018, https://ihsmarkit .com/products/polyester-fibers-chemical-economics-handbook.html.

11. Hermes, *Enough for One Lifetime*, 291.

12. "Vogue Presents Fashions of the Future," *Vogue*, February 1, 1939, 71–81, 137–146; "Clothing of the Future—Clothing in the Year 2000," Pathetone Weekly, YouTube video, 1:26, www.youtube.com/watch?v=U9eAiy0IGBI.

13. Regina Lee Blaszczyk, "Styling Synthetics: DuPont's Marketing of Fabrics and Fashions in Postwar America," *Business History Review*, Autumn 2006, 485–528; Ronald Alsop, "Du Pont Acts to Iron Out the Wrinkles in Polyester's Image," *Wall Street Journal*, March 2, 1982, 1.

14. Jean E. Palmieri, "Under Armour Scores $1 Billion in Sales through Laser Focus on Athletes," *WWD*, December 1, 2011, https://wwd.com/wwd-publications/wwd-special-report/2011-12-01-2104533/; Jean E. Palmieri, "Innovating the Under Armour Way," *WWD*, August 10, 2016, 11–12; Kelefa Sanneh, "Skin in the Game," *New Yorker*, March 24, 2014, www.newyorker.com/magazine/2014/03/24/skin-in-the-game.

15. Phil Brown, interview with the author, March 4, 2015; Virginia Postrel, "How the Easter Bunny Got So Soft," Bloomberg Opinion, April 2, 2015, https://vpostrel.com/articles/how-the-easter-bunny-got-so-soft.

16. Brian K. McFarlin, Andrea L. Henning, and Adam S. Venable, "Clothing Woven with Titanium Dioxide-Infused Yarn: Potential to Increase Exercise Capacity in a Hot, Humid Environment?" *Journal of the Textile Institute* 108 (July 2017): 1259–1263, https://doi.org/10.1080/00405000.2016.1239329.

17. Elizabeth Miller, "Is DWR Yucking Up the Planet?" SNEWS, May 12, 2017, www.snewsnet.com/news/is-dwr-yucking-up-the-planet; John Mowbray, "Gore PFC Challenge Tougher than Expected," *EcoTextile News*, February 20, 2019, www.ecotextile.com/2019022024078/dyes-chemicals-news/gore-pfc-challenge-tougher-than-expected.html. Whether the compounds actually pose significant risks is controversial. But as a consumer brand, Under Armour doesn't need to adjudicate those claims any more than it has to decide whether blue or red is a better color. Its job is to satisfy customers.

18. Kyle Blakely, interview with the author, July 31, 2019.

19. Christian Holland, "MassDevice Q&A: OmniGuide Chairman Yoel Fink," MassDevice, June 1, 2010, www.massdevice.com/massdevice-qa-omniguide-chairman-yoel-fink/; Bruce Schechter, "M.I.T. Scientists Turn Simple Idea Into 'Perfect Mirror,'" *New York Times*, December 15, 1998, sec. F, 2, www.nytimes.com/1998/12/15/science/mit-scientists-turn-simple-idea-into-perfect-mirror.html.

20. Yoel Fink, interviews with the author, July 28, 2019, and August 16, 2019; Bob D'Amelio and Tosha Hays, interviews with the author, July 29, 2019, and August 28, 2019; Jonathon Keats, "This Materials Scientist Is on a Quest to Create Functional Fibers That Could Change the Future of Fabric," *Discover*, April 2018, http://discovermagazine.com/2018/apr/future-wear; David L. Chandler, "AFFOA Launches State-of-the-Art Facility for Prototyping Advanced Fabrics," MIT News Office, June 19, 2017, https://news.mit.edu/2017/affoa-launches-state-art-facility-protoyping-advanced-fabrics-0619. Fink and Hays left AFFOA in late 2019, but Fink's MIT group continues to research functional fibers.

21. Fink was born in the United States, but his family emigrated when he was two.

22. Hiroyasu Furukawa, Kyle E. Cordova, Michael O'Keeffe, and Omar M. Yaghi, "The Chemistry and Applications of Metal-Organic Frameworks," *Science* 341, no. 6149 (August 30, 2013): 974.

23. Juan Hinestroza, interviews with the author, August 23, 2019, August 30, 2019, and September 3, 2019, and emails to the author September 2, 2019, September 5, 2019, and September 25, 2019; College of Textiles, NC State University, "Researchers Develop High-Tech, Chemical-Resistant Textile Layers," *Wolftext*, Summer 2005, 2, https://sites.textiles.ncsu.edu/wolftext-alumni-newsletter/wp-content/uploads/sites/53/2012/07/wolftextsummer2005.pdf; Ali K. Yetisen, Hang Qu, Amir Manbachi, Haider Butt, Mehmet R. Dokmeci, Juan P. Hinestroza, Maksim Skorobogatiy, Ali Khademhosseini, and SeokHyun Yun, "Nanotechnology in Textiles," *ACS Nano*, March 22, 2016, 3042–3068.

24. In 2016, Allergan sold the technology to Sofregen Medical, another silk-oriented medical spinoff from Tufts. Sarah Faulkner, "Sofregen Buys Allergan's Seri Surgical Scaffold," MassDevice, November 14, 2016, www.massdevice.com/sofregen-buys-allergans-seri-surgical-scaffold/.

25. Rachel Brown, "Science in a Clean Skincare Direction," *Beauty Independent*, December 6, 2017, www.beautyindependent.com/silk-therapeutics/.

26. Benedetto Marelli, Mark A. Brenckle, and David L. Kaplan, "Silk Fibroin as Edible Coating for Perishable Food Preservation," *Science Reports* 6 (May 6, 2016): art. 25263, www.nature.com/articles/srep25263.

27. Kim Bhasin, "Chanel Bets on Liquid Silk for Planet-Friendly Luxury," Bloomberg, June 11, 2019, www.bloomberg.com/news/articles/2019-06-11/luxury-house-chanel-takes-a-minority-stake-green-silk-maker.

28. Department of Energy Advanced Research Projects Agency (ARPA-E), "Personal Thermal Management to Reduce Energy Consumption Workshop," https://arpa-e.energy.gov/?q=events/personal-thermal-management-reduce-building-energy-consumption-workshop.

29. Centre for Industry Education Collaboration, University of York, "Poly(ethene) (Polyethylene)," *Essential Chemical Industry (ECI)—Online*, www.essentialchemicalindustry.org/polymers/polyethene.html; Svetlana V. Boriskina, "An Ode to Polyethylene," *MRS Energy & Sustainability* 6 (September 19, 2019), https://doi.org/10.1557/mre.2019.15.

30. Svetlana Boriskina, interview with the author, July 30, 2019, and emails to the author, August 15, 2019, and September 2, 2019.

INDEX

This index does not include the references. For a searchable reference list, see https://vpostrel.com/the-fabric-of-civilization/references.

Page numbers followed by n or nn indicate notes. Artwork and figures are indicated by page number and "(fig.)"; photographs are indicated by page number and "(photo)"; tables are indicated by page number and "(table)."

Virginia Postrel is an award-winning journalist and independent scholar. A columnist for *Bloomberg Opinion*, she previously served as a columnist for the *Wall Street Journal*, *The Atlantic*, the *New York Times*, and *Forbes*. She is the author of the highly acclaimed *The Substance of Style*, *The Power of Glamour*, and *The Future and Its Enemies*. Her research is supported by the Alfred P. Sloan Foundation. She lives in Los Angeles, California.